Miller's Guide to Foundations & Sitework

Home Reference Series

Miller's Guide to Foundations & Sitework

REX MILLER
Professor Emeritus
State University College at Buffalo
Buffalo, New York

MARK R. MILLER
Professor
The University of Texas at Tyler
Tyler, Texas

McGraw-Hill

New York   Chicago   San Francisco   Lisbon   London
Madrid   Mexico City   Milan   New Delhi   San Juan
Seoul   Singapore   Sydney   Toronto

Library of Congress Cataloging-in-Publication Data.

Miller, Rex.
    Miller's guide to foundations and sitework / Mark R. Miller, Rex Miller.
    p.   cm.
    Includes index.
    ISBN 0-07-145145-5
    1. Dwellings—Foundations. 2. Building sites. I. Title: Guide to foundations and
sitework. II. Miller, Mark. III. Title.

    TH2101.M54 2005
    690'.8—dc22

                     2005047866

1 2 3 4 5 6 7 8 9 0   QPD/QPD   0 1 0 9 8 7 6 5

ISBN 0-07-145145-5

*The sponsoring editor for this book was Larry Hager, the editing supervisor was
Caroline Levine, and the production supervisor was Sherri Souffrance. The art director for
the cover was Handel Low. It was set in ITC Century Light by Kim J. Sheran of McGraw-
Hill Professional's Hightstown, N.J., composition unit.*

McGraw-Hill books are available at special quantity discounts to use as premiums and sales
promotions, or for use in corporate training programs. For more information, please write to
the Director of Special Sales, McGraw-Hill Professional, Two Penn Plaza, New York, NY
10121. Or contact your local bookstore.

Contents

4 Pouring Concrete Slabs & Floors

5 Building Floor Frames

6 Private Sewage Facilities

7 Private Water Systems

8 Designing and Planning for Solar Heating

9 Alternative Types of Foundations

Appendices

Glossary

Preface

Miller's Guide to Foundations & Sitework is written for everyone who wants or needs to know about carpentry and construction. Whether you are remodeling an existing home or building a new one, the rewards from a job well done are many.

This text can be used by students in vocational courses, technical colleges, apprenticeship programs, and construction classes in industrial technology programs. The home do-it-yourselfer will find answers to many questions that pop up in the course of getting a job done whether over a weekend or over a year's time.

To prepare this text, the authors examined courses of study in schools located all over the country. An effort was made to take into consideration the geographic differences and special environmental factors relevant to a particular area. For instance, some loca-tions are not suited for the basement type of house. The slab method of construction for a house foundation is also shown with its many ramifications and variations. Whether it is a basement or slab type of house, it is important that the proper procedures and materials are used under the right circumstances in order to produce living quarters suitable for habitation over a period of 50 to 100 years.

No book can be completed without the aid of many people. The Acknowledgments that follow mention some of those who contributed to making this text the most current in design and technology techniques available to the carpenter. We trust you will enjoy using the book as much as we enjoyed writing it.

REX MILLER
MARK R. MILLER

Acknowledgments

Miller's Guide to Foundations & Sitework has been prepared with the aid of many people. The acknowledgments that follow mention some of those who contributed to making this text the most current in design and technology techniques available to the carpenter. We trust you will enjoy using the book as much as we enjoyed writing it.

The authors would like to thank the following manufacturers for their generous support. They furnished photographs, drawings, and technical assistance. Without the donations of time and effort on the part of many people, this book would not have been possible. We hope this acknowledgment of some of the contributions will let you know that the field you are working in or are about to enter is one of the best.

American Plywood Association
Anderson Windows
Black and Decker
David White Instruments
DeWalt
Diamond Well Drilling
Duo Fast Corp.
Flexcon Industries
Forest Products Laboratory
Fox and Jacobs
Georgia Pacific
IRL Daffin
Jet, Inc.
Millers Falls, Division of Ingersol-Rand Co.
National Terrazzo and Mosaic Association
Portland Cement Association
Power Tools Division, Rockwell International
Proctor Products
Rhodes Drilling
Rubber Polymer Corp.
Skrobarczk Properties
Stabilia
Stanley Tools
Universal Form Clamp Co.

Introduction

Houses do not exist in isolation. Conceived for the purpose of supporting a wide range of human activities, houses are built in response to sociocultural, economic, and political needs. They are erected in natural and planned areas that constrain as well as offer opportunities for development. That means we should carefully consider the environmental forces that a house site represents.

Topography, plants, and the microclimate of the site also influence design decisions. This planning should be done at a very early stage in the design process. Human comfort as well as energy conservation and resources should be considered. All aspects of the area should figure into the design.

The indigenous qualities of a place should be adaptable to the building requirements. Landscaping should take into account the following:

- Path of the sun
- Direction of the prevailing wind
- Natural flow of water on a site

In addition to environmental forces, there are always the regulatory forces of zoning ordinances. These regulations prescribe the acceptable uses and activities for a building site as well as limit the size and shape of the building itself and where it may be located on the site.

Just as environmental and regulatory factors influence where and how a building is constructed, the construction and use of a building inevitably place a demand on local systems, utilities, and other services. A fundamental question is, How much construction can a site sustain without exceeding the capacity of these services or causing harmful effects on the environment? In addition to altering land use, the construction of a building affects the environment by utilizing energy and consuming materials. Building only as much as needed is a necessary first step in reducing the amount of resources needed for construction.

Consideration of these forces as well as site planning elements that modify a site for access and use begins with a careful analysis and takes into consideration prior and existing building codes.

BUILDING CODES
History

There were three major model codes:

- The *National Building Code* was developed and published by Building Officials and Code Administrators International, Inc. (BOAC) and was used primarily in the northeastern United States.
- The *Uniform Building Code* (UBC) was developed and published by the International Conference of Building Officials (ICBO) and used primarily in the central and western United States.
- The *Standard Building Code* (SBC) was developed and published by the Southern Building Code Conference (SBCC) and used primarily in the southeastern United States.

A new *International Building Code* (IBC) was developed by the International Code Council (ICC). It was first published in 2000. This is the first unified model code in U.S. history; it is uncertain how long it will take this code to be fully adopted.

Check with the local authority—city, county, or state—to be sure which code is used by the building inspectors associated with the permit-granting authority.

Other Codes

The code used by the local electrical inspector almost always is the *National Electrical Code,* which is published by the National Fire Protection Association (NFPA) to ensure the safety of persons and the safeguarding of buildings and their contents from hazards arising from the use of electricity for heat, light, and power.

The *Life Safety Code,* also published by the NFPA, establishes minimum requirements for fire safety; the prevention of danger from fire, smoke, and gases; fire detection and alarm systems; fire extinguishing systems; and emergency egress.

The *Safety Code for Elevators and Escalators* is published by the American National Standards Institute. This code comes into play with multistoried housing with elevators.

1
CHAPTER

Starting the Job

BECAUSE CARPENTRY INVOLVES ALL KINDS of challenging jobs, it is an exciting industry. You will have to work with hand tools, power tools, and all types of building materials. You can become very skilled at your job. You get a chance to be proud of what you do. You can stand back and look at the building you just helped erect and feel great about a job well done.

One of the exciting things about being a carpenter is watching a building come up. You actually see it grow from the ground up. Many people work with you to make it possible to complete the structure. Being part of a team can be rewarding, too.

This book will help you do a good job in carpentry, starting from the ground up. Because it covers all the basic construction techniques it will aid you in making the right decisions in your choice of house site, use of the proper footings, and checking the foundation for soundness of construction. More information is available in the other five books of this series.

You have to do something over and over again to gain skill. When reading this book, you might not always get the idea the first time. Go over it again until you understand. Then go out and practice what you just read. This way you can see for yourself how the instructions actually work. Of course, no one can learn carpentry by merely reading a book. You have to read, reread, and then do. This "do" part is the most important. You have to take the hammer or saw in hand and actually do the work. There is nothing like good, honest sweat from a hard day's work. At the end of the day you can say "I did that" and be proud that you did.

This chapter should help you build these skills:

- Select personal protective gear
- Work safely as a carpenter
- Measure building materials
- Lay out building parts
- Cut building materials
- Fasten materials
- Shape and smooth materials
- Identify basic hand tools
- Recognize common power tools

SAFETY

Figure 1-1 shows a carpenter using one of the latest means of driving nails: the compressed-air-driven nail driver, which drives nails into the wood with a single stroke. The black cartridge that appears to run up near

Fig. 1-1 *This carpenter is using an air-driven nail driver to nail these framing members.* (Duo-Fast)

the carpenter's leg is a part of the nailer. It holds the nails and feeds them as needed.

As for safety, notice the carpenter's shoes. They have rubber soles for gripping the wood. This will prevent a slip through the joists and a serious fall. The steel toes in the shoes prevent damage to the foot from falling materials. The soles of the shoes are very thick to prevent nails from going through. The hard hat protects the carpenter's head from falling lumber, shingles, or other building materials. The carpenter's safety glasses cannot be seen in Fig. 1-1, but they are required equipment for the safe worker.

Other Safety Measures

To protect the eyes, it is best to wear safety glasses. Make sure your safety glasses are of tempered glass. They will not shatter and cause eye damage. In some instances you should wear goggles. This prevents splinters and other flying objects from entering the eye from under or around the safety glasses. Ordinary glasses aren't always the best, even if they are tempered glass. Just become aware of the possibilities of

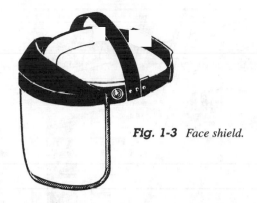

Fig. 1-3 *Face shield.*

Fig. 1-2 *Safety glasses.*

eye damage whenever you start a new job or procedure. See Fig. 1-2 for a couple of types of safety glasses.

Sneakers are used only by roofers. Sneakers, sandals, and dress shoes do not provide enough protection for the carpenter on the job. Only safety shoes should be worn on the job.

Gloves Some types of carpentry work require the sensitivity of the bare fingers. Other types do not require the hands or fingers to be exposed. In cold or even cool weather, gloves may be in order. Gloves are often needed to protect your hands from splinters and rough materials. It's only common sense to use gloves when handling rough materials.

Probably the best gloves for carpenter work are a lightweight type. A suede finish to the leather improves the gripping ability of the gloves. Cloth gloves tend to catch on rough building materials. They may be preferred, however, if you work with short nails or other small objects.

Body protection Before you go to work on any job, make sure your entire body is properly protected. The hard hat comes in a couple of styles. Under some conditions the face shield is better protection. See Fig. 1-3.

Is your body covered with heavy work clothing? This is the first question to ask before going onto the job site. Has as much of your body as practical been covered with clothing? Has your head been properly protected? Are your eyes covered with approved safety glasses or face shield? Are your shoes sturdy, with

safety toes and steel soles to protect against nails? Are gloves available when you need them?

General Safety Rules

Some safety procedures should be followed at all times. This applies to carpentry work especially:

- Pay close attention to what is being done.
- Move carefully when walking or climbing.
- (Take a look at Fig. 1-4. This type of made-on-the-job ladder can cause trouble.) Use the leg muscles when lifting.
- Move long objects carefully. The end of a carelessly handled 2 × 4 can damage hundreds of dollars' worth of glass doors and windows. Keep the workplace neat and tidy. Figure 1-5A shows a cluttered working area. It would be hard to walk along here without tripping. If a dumpster is used for trash and debris, as in Fig. 1-5B, many accidents can be prevented. Sharpen or replace dull tools.
- Disconnect power tools before adjusting them.
- Keep power tool guards in place.
- Avoid interrupting another person who is using a power tool.
- Remove hazards as soon as they are noticed.

Safety on the Job

A safe working site makes it easier to get the job done. Lost time due to accidents puts a building schedule behind. This can cost many thousands of dollars and lead to late delivery of the building. If the job is properly organized and safety is taken into consideration, the smooth flow of work is quickly noticed. No one wants to get hurt. Pain is no fun. Safety is just common sense. If you know how to do something safely, it will not take any longer than if you did it in an unsafe manner. Besides, why would you deliberately do something that is dangerous? All that safety requires is a

Fig. 1-4 *A made-on-the-job ladder.*

(A)

(B)

Fig. 1-5 *(A) Cluttered work site. (B) A work area can be kept clean if a large dumpster is kept nearby for trash and debris.*

few precautions on the job. Safety becomes a habit once you get the proper attitude established in your thinking. Some of these important habits to acquire are as follows:

- Know exactly what is to be done before you start a job.
- Use a tool only when it can be used safely. Wear all safety clothing recommended for the job. Provide a safe place to stand to do the work. Set ladders securely. Provide strong scaffolding.
- Avoid wet, slippery areas.
- Keep the working area as neat as practical.
- Remove or correct safety hazards as soon as they are noticed. Bend protruding nails over. Remove loose boards.
- Remember where other workers are and what they are doing.
- Keep fingers and hands away from cutting edges at all times.
- Stay alert!

Safety Hazards

Carpenters work in unfinished surroundings. While a house is being built, there are many unsafe places around the building site. You have to stand on or climb ladders, which can be unsafe. You may not have a good footing while standing on a ladder. You may not be climbing a ladder in the proper way. Holding onto the rungs of the ladder is very unsafe. You should always hold onto the outside rails of the ladder when climbing.

There are holes that can cause you to trip. They may be located in the front yard where the water or sewage lines come into the building. There may be holes for any number of reasons. These holes can cause you all kinds of problems, especially if you fall into them or turn your ankle.

The house in Fig. 1-6 is almost completed. However, if you look closely, you can see that some wood has been left on the garage roof. This wood can slide down and hit a person working below. The front porch has not been poured. This means stepping out of the front door can be a rather long step. Other debris around the yard can be a

Fig. 1-6 *Even when a house is almost finished, there can still be hazards. Wood left on a roof could slide off and hurt someone, and without the front porch, it is a long step down.*

source of trouble. Long slivers of flashing can cause trouble if you step on them and they rake your leg. You have to watch your every step around a construction site.

Outdoor work Much of the time carpentry is performed outdoors. This means you will be exposed to the weather, so dress accordingly. Wet weather increases the accident rate. Mud can make it hard to find a secure place to stand. Mud can also cause you to slip if you don't clean it off your shoes. Be very careful when it is muddy and you are climbing on a roof or a ladder.

Tools Any tool that can cut wood can cut flesh. You have to keep in mind that although tools are an aid to the carpenter, they can also be a source of injury. A chisel can cut your hand as easily as the wood. In fact, it can do a quicker job on your hand than on the wood it was intended for. Saws can cut wood and bones. Be careful with all types of saws, both hand and electric. Hammers can do a beautiful job on your fingers if you miss the nail head. The pain involved is intensified in cold weather. Broken bones can be easily avoided if you keep your eye on the nail while you're hammering. Besides that, you will get the job done more quickly. And, after all, that's why you are there—to get the job done and do it right the first time. Tools can help you do the job right. They can also cause you injury. The choice is up to you.

In order to work safely with tools you should know what they can do and how they do it. The next few pages are designed to help you use tools properly.

USING CARPENTER TOOLS

A carpenter is lost without tools. This means you have to have some way of containing them. A toolbox is very important. If you have a place to put everything,

then you can find the right tool when you need it. A toolbox should have all the tools mentioned here. In fact, you will probably add more as you become more experienced. Tools have been designed for every task. All it takes is a few minutes with a hardware manufacturer's catalog to find just about everything you'll ever need. If you can't find what you need, the manufacturers are interested in making it.

Measuring Tools

Folding rule When you are using the folding rule, place it flat on the work. The 0 end of the rule should be exactly even with the end of the space or board to be measured. The correct distance is indicated by the reading on the rule.

A very accurate reading may be obtained by turning the edge of the rule toward the work. In this position, the marked gradations of the face of the rule touch the surface of the board. With a sharp pencil, mark the exact distance desired. Start the mark with the point of the pencil in contact with the mark on the rule. Move the pencil directly away from the rule while making the mark.

One problem with the folding rule is that it breaks easily if it is twisted. This happens most commonly when it is being folded or unfolded. The user may not be aware of the twisting action at the time. You should keep the joints oiled lightly. This makes the rule operate more easily.

Pocket tape Beginners may find the pocket tape (Fig. 1-7) the most useful measuring tool for all types of work. It extends smoothly to full length. It returns quickly to its compact case when the return button is pressed. Steel tapes are available in a variety of lengths. For most carpentry a rule 6, 8, 10, or 12 feet long is used.

Longer tapes are available. They come in 20-, 50-, and 100-foot lengths. See Fig. 1-8. This tape can be extended to 50 feet to measure lot size and the location of a house

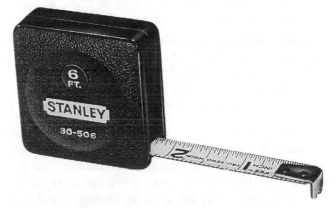

Fig. 1-7 *Tape measure.* (Stanley Tools)

Fig. 1-8 *A longer tape measure.* (Stanley Tools)

on a lot. It has many uses around a building site. A crank handle can be used to wind it up once you are finished with it. The hook on the end of the tape makes it easy for one person to use it. Just hook the tape over the end of a board or nail and extend it to the desired length.

Saws

Carpenters use a number of different saws. These saws are designed for specific types of work. Many are misused. They will still do the job, but they would do a better job if used properly. Handsaws take quite a bit of abuse on a construction site. It is best to buy a good-quality saw and keep it lightly oiled.

Standard skew-handsaw This saw has a wooden handle. It has a 22-inch length. A 10-point saw (with 10 teeth per inch) is suggested for crosscutting. Crosscutting means cutting wood *across* the grain. The 26-inch-length, 5½-point saw is suggested for ripping, or cutting *with* the wood grain.

Figure 1-9 shows a carpenter using a handsaw. This saw is used in places where the electric saw cannot be used. Keeping it sharp makes a difference in the quality of the cut and the ease with which it can be used.

Backsaw The backsaw gets its name from the piece of heavy metal that makes up the top edge of the cutting part of the saw. See Fig. 1-10. It has a fine-tooth configuration. This means it can be used to cut cross-grain and leave a smoother finished piece of work. This type of saw is used by finish carpenters who want to cut trim or molding.

Miter box As you can see from Fig. 1-11A, the miter box has a backsaw mounted in it. This box can be adjusted using the lever under the saw handle. You can adjust it for the cut you wish. It can cut from 90° to 45°. It is used for finish cuts on moldings and trim materials. The angle of the cut is determined by the location of the saw in reference to the bed of the box. Release the clamp on the bottom of the saw support to adjust the saw to any degree desired. The wood is held with one hand against the fence of the box and the bed. Then the saw is used by the other hand. As you can see from the setup, the cutting should take place when the saw is pushed forward. The backward movement of the saw should be made with the pressure on the saw released slightly. If you try to cut on the backward movement, you will just pull the wood away from the fence and damage the quality of the cut.

Powered compound miter saw In most cases the miter box has been replaced by the powered miter saw. It is portable, inexpensive, and, in most instances, easily mastered by the do-it-yourselfer. Greater accuracy is attained in the cuts and fits by using the power saw because the opportunity for movement of the wood being cut is not so prevalent. Just use caution when operating the saw. It can be very dangerous if used carelessly. (See Fig. 1-11B.)

Coping saw Another type of saw the carpenter can make use of is the coping saw (Fig. 1-12). This one can cut small thicknesses of wood at any curve or angle desired. It can be used to make sure a piece of paneling fits properly or a piece of molding fits another piece in the corner. The blade is placed in the frame with the teeth pointing toward the handle. This means it cuts only on the downward stroke. Make sure you properly support the piece of wood being cut. A number of blades can be obtained for this type of saw. The number of teeth in the blade determines the smoothness of the cut.

Hammers and Other Small Tools

There are a number of different types of hammers. The one the carpenter uses is the *claw* hammer. It has claws that can extract nails from wood if the nails have been put in the wrong place or have bent while being driven. Hammers can be bought in 20-ounce, 24-ounce, 28-ounce, and 32-ounce weights for carpentry work. Most carpenters prefer a 20-ounce. You have to work with a number of different weights to find out which will work best for you. Keep in mind that the hammer should be of tempered steel. If the end of the hammer has a tendency to splinter or chip off when it hits a nail, the pieces can hit you in the eye or elsewhere, causing serious damage. It is best to wear safety glasses whenever you use a hammer.

Nails are driven by hammers. Figure 1-13 shows the gage, inch, and penny relationships for the common box nail. The *d* after the number means *penny*. This is a measuring unit inherited from the English in the colonial days. There is little or no relationship between

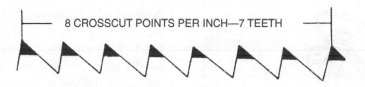

8 CROSSCUT POINTS PER INCH—7 TEETH

6 RIP POINTS PER INCH—5 TEETH

THE NUMBER OF POINTS PER INCH ON A HANDSAW
DETERMINES THE FINENESS OR COARSENESS
OF CUT. MORE POINTS PRODUCE A FINER CUT.

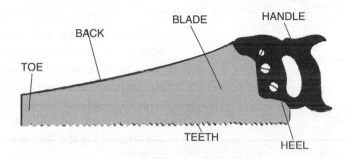

BACK   BLADE   HANDLE

TOE

TEETH   HEEL

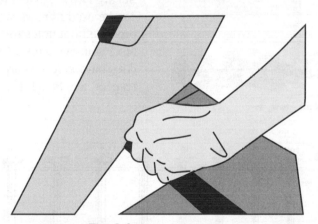

Fig. 1-9 *Using a handsaw.*

Fig. 1-10 *Backsaw.* (Stanley Tools)

penny and inches. If you want to be able to talk about it intelligently, you'll have to learn both inches and penny. The gage is nothing more than the American Wire Gage number for the wire that the nails were made from originally. Finish nails have the same measuring unit (penny) but do not have the large, flat heads.

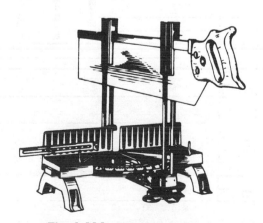

Fig. 1-11A *Miter box.* (Stanley Tools)

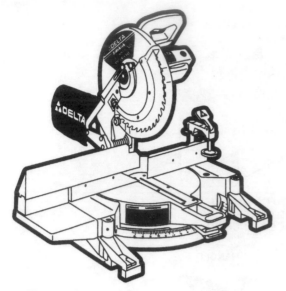

Fig. 1-11B *Powered compound miter saw.* (Delta)

Fig. 1-12 *Coping saw.* (Stanley Tools)

Nail set Finish nails are driven below the surface of the wood by a nail set. The nail set is placed on the head of the nail. The large end of the nail set is struck by the hammer. This causes the nail to go below the surface of the wood. Then the hole left by the counter-sunk nail is filled with wood filler and finished off with a smooth coat of varnish or paint. Figure 1-14 shows the nail set and its use.

The carpenter would be lost without a hammer. See Fig. 1-15. Here the carpenter is placing sheathing on rafters to form a roof base. The hammer is used to drive the boards into place, since they have to overlap slightly. Then the nails are driven by the hammer also.

In some cases a hammer will not do the job. The job may require a hatchet. See Fig. 1-16. This device can be used to pry and to drive. It can pry boards loose when they are improperly installed. It can sharpen posts to be driven at the site. The hatchet can sharpen the ends of stakes for staking out the site. It can also withdraw nails. This type of tool can also be used to drive stubborn sections of a wall into place when they are erected for the first time. The tool has many uses.

Scratch awl An awl is a handy tool for a carpenter. It can be used to mark wood with a scratch mark and to produce pilot holes for screws. Once it is in your tool-box, you can think of a hundred uses for it. Since it does have a very sharp point, it is best to treat it with respect. See Fig. 1-17.

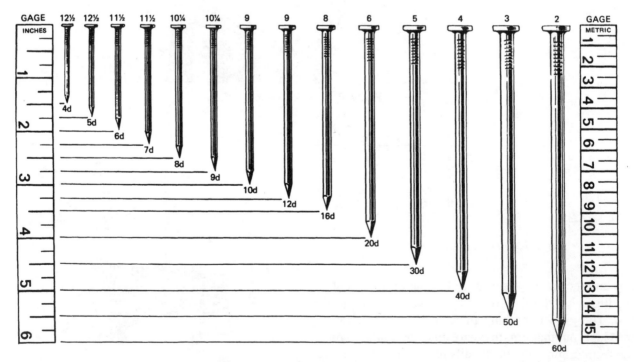

Fig. 1-13 *Nails.* (Forest Products Laboratory)

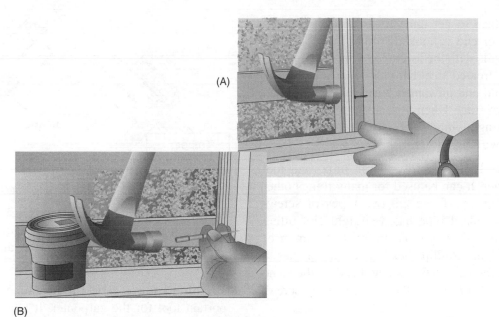

(A)

(B)

Fig. 1-14 *(A) Driving a nail with a hammer. (B) Finishing the job with a nail set to make sure the hammer doesn't leave an impression in the soft wood of the window frame.*

Fig. 1-15 *Putting on roof sheathing. The carpenter is using a hammer to drive the board into place.*

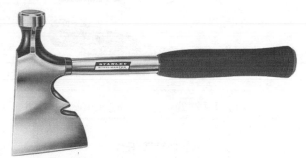

Fig. 1-16 *Hatchet. (Stanley Tools)*

Fig. 1-17 *Scratch awl. (Stanley Tools)*

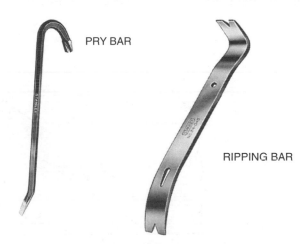

PRY BAR

RIPPING BAR

Fig. 1-18 *Wrecking bars. (Stanley Tools)*

Wrecking bar This device (Fig. 1-18) has a couple of names, depending on which part of the country you are in at the time. It is called a wrecking bar in some parts and a crowbar in others. One end has a chisel-sharp flat surface to get under boards and pry them loose. The other end is hooked so that the slot in the end can pull nails with the leverage of the long handle. This specially treated steel bar can be very helpful in prying away old and unwanted boards. It can be used to help give leverage when you are putting a wall in place and making it plumb. This tool has many uses for the carpenter with ingenuity.

Screwdrivers The screwdriver is an important tool for the carpenter. It can be used for many things other than turning screws. There are two types of screwdrivers. The standard type has a straight slot-fitting blade at its end. This type is the most common of screwdrivers. The Phillips-head screwdriver has a cross or X on the end to fit a screw head of the same design. Figure 1-19 shows the two types of screwdrivers.

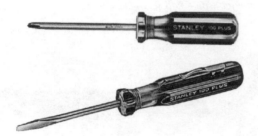

Fig. 1-19 *Two types of screwdrivers.*

Squares

In order to make corners meet and standard sizes of materials fit properly, you must have things square. That calls for a number of squares to check that the two walls or two pieces come together at a perpendicular.

Try square The *try square* can be used to mark small pieces for cutting. If one edge is straight and the handle part of the square (Fig. 1-20) is placed against this straight edge, then the blade can be used to mark

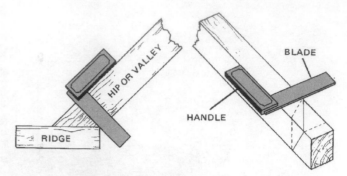

Fig. 1-20 *Use of a try square.* (Stanley Tools)

the wood perpendicular to the edge. This comes in handy when you are cutting 2 × 4s and want them to be square.

Framing square The framing square is a very important tool for the carpenter. It allows you to make square cuts in dimensional lumber. This tool can be used to lay out rafters and roof framing. See Fig. 1-21. It is also used to lay out stair steps.

Later in this book you will see a step-by-step procedure for using the framing square. The tools are described as they are called for in actual use.

Bevel A bevel can be adjusted to any angle to make cuts at the same number of degrees. See Fig. 1-22. Note how the blade can be adjusted. Now take a look at Fig. 1-23. Here you can see the overhang of rafters. If you want the ends to be parallel with the side of the house, you can use the bevel to mark them before they are cut off. Simply adjust the bevel so the handle is on top of the rafter and the blade fits against the soleplate below. Tighten the screw and move the bevel down the rafter to where you want the cut. Mark the angle along the blade of the bevel. Cut along the mark, and you have what you see in Fig. 1-23. It is a good device for transferring angles from one place to another.

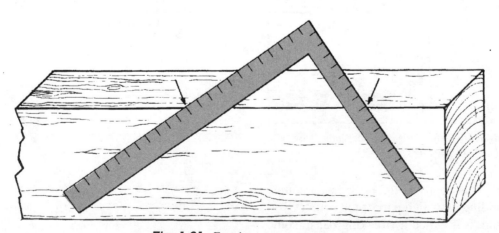

Fig. 1-21 *Framing square.* (Stanley Tools)

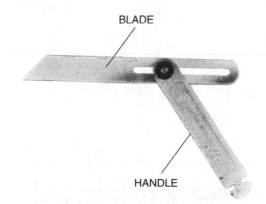

BLADE

HANDLE

Fig. 1-22 *Bevel.* (Stanley Tools)

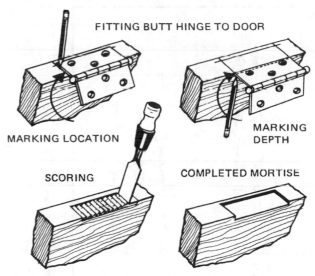

FITTING BUTT HINGE TO DOOR

MARKING LOCATION

MARKING DEPTH

SCORING

COMPLETED MORTISE

Fig. 1-24 *Using a wood chisel to complete a mortise.*

Fig. 1-23 *Rafter overhang cut to a given angle.*

on the handle so the force of the hammer blows will not chip the handle. Other applications are up to you, the carpenter. You'll find many uses for the chisel in making things fit.

Plane Planes (Fig. 1-25) are designed to remove small shavings of wood along a surface. One hand holds the knob in front, and the other holds the handle in back. The blade is adjusted so that only a small sliver of wood is removed each time the plane is passed over the wood. It can be used to make sure that doors and windows fit properly. It can be used for any number of wood smoothing operations.

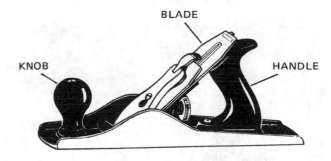

BLADE

KNOB

HANDLE

Fig. 1-25 *Smooth plane.* (Stanley Tools)

Chisel Occasionally you may need a wood chisel. It is sharpened on one end. When the other end is struck with a hammer, the cutting end will do its job. That is so, of course, if you have kept it sharpened. See Fig. 1-24.

The chisel is commonly used in fitting or hanging doors. It is used to remove the area where the door hinge fits. Note how it is used to score the area (Fig. 1-24); it is then used at an angle to remove the ridges. A great deal of the work with the chisel is done by using the palm of the hand as the force behind the cutting edge. A hammer can be used. In fact, chisels have a metal tip

Dividers and compass Occasionally a carpenter must draw a circle. This is done with a compass. The compass shown in Fig. 1-26A can be converted to a divider by removing the pencil and inserting a straight steel pin. The compass has a sharp point that fits into the wood surface. The pencil part is used to mark the circle's circumference. It is adjustable to various radii.

The dividers in Fig. 1-26A have two points made of hardened metal. They are adjustable. It is possible to use them to transfer a given measurement from the

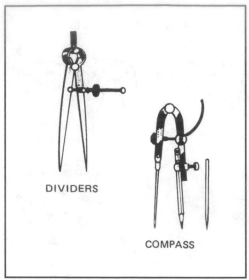

Fig. 1-26A *Dividers and compass*

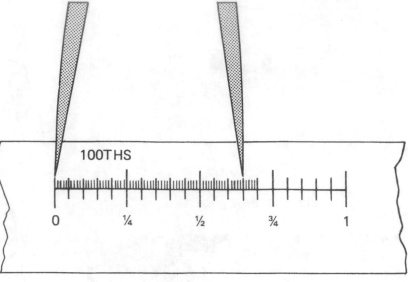

Fig. 1-26B *Dividers being used to transfer hundredths of an inch.*

framing square or measuring device to another location. See Fig. 1-26B.

Level In order to have things look as they should, a level is necessary. There are a number of sizes and shapes available. The one shown in Figure 1-27B is the most common type used by carpenters. The bubbles in the glass tubes tell you if the level is obtained. In Fig. 1-27A the carpenter is using the level to make sure a window is in properly before nailing it into place permanently.

If the vertical and horizontal bubbles are lined up between the lines, then the window is plumb, or verti-

HORIZONTAL LEVEL INDICATOR

VERTICAL INDICATOR

VERTICAL INDICATOR

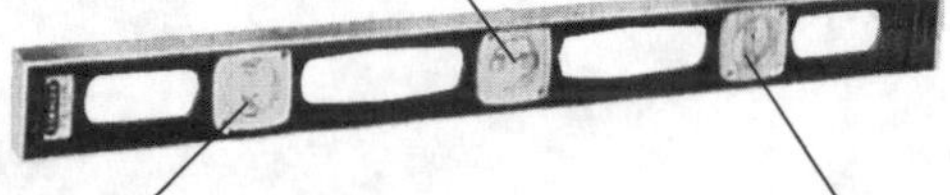

Fig. 1-27B *A commonly used type of level.* (Stanley Tools)

cal. A plumb bob is a small, pointed weight. It is attached to a string and dropped from a height. If the bob is just above the ground, it will indicate the vertical direction by its string. Keeping windows, doors, and frames square and level makes a difference in fitting. It is much easier to fit prehung doors into a frame that is square. When it comes to placing panels of 4- × 8-foot plywood sheathing on a roof or on walls, squareness can make a difference as to fit. Besides, a square fit and a plumb door and window look better than those that are a little off. Figure 1-27C shows three plumb bobs.

Fig. 1-27A *Using a level to make sure a window is placed properly before nailing.* (Andersen)

Fig. 1-27C *Plumb bobs.* (Stanley Tools)

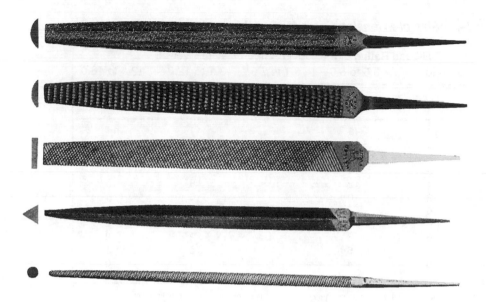

Fig. 1-28 Wood and cabinet files: (A) Half-round; (B) Rasp; (C) Flat; (D) Triangular; and (E) Round. (Millers Falls Division, a division of Ingersol-Rand Co.)

Files A carpenter finds use for a number of types of files. The files have different surfaces for doing different jobs. Tapping out a hole to get something to fit may be just the job for a file. Some files are used for sharpening saws and touching up tool cutting edges. Figure 1-28 shows different types of files. Other files may also be useful. You can acquire them later as you develop a need for them.

Clamps C-clamps are used for many holding jobs. They come in handy when you are placing kitchen cabinets by holding them in place until screws can be inserted and properly seated. This type of clamp can be used for an extra hand every now and then when two hands aren't enough to hold a combination of pieces until you can nail them. See Fig. 1-29.

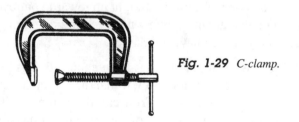

Fig. 1-29 C-clamp.

Cold chisel It is always good to have a cold chisel around. It is very much needed when you can't remove a nail. Its head may have broken off and the nail must be removed. The chisel can cut the nail and permit the separation of the wood pieces. See Fig. 1-30.

Fig. 1-30 Cold chisel. (Stanley Tools)

If a chisel of this type starts to "mushroom" at the head, you should remove the splintered ends with a grinder. Hammering on the end can produce a mushrooming effect. These pieces should be taken off since they can easily fly off when hit with a hammer. That is another reason for using eye protection when you are using tools.

Caulking gun In times of energy crisis, the caulking gun gets plenty of use. It is used to fill in around windows and doors and everywhere there may be an air leak. There are many types of caulk being made today.

This gun is easily operated. Insert the cartridge and cut its tip to the shape you want. Puncture the thin plastic film inside. A bit of pressure will cause the caulk to come out the end. The long rod protruding from the end of the gun is turned over. This is done so the serrated edge will engage the hand trigger. Remove the pressure from the cartridge when you are finished. Do this by rotating the rod so that the serrations are not engaged by the trigger of the gun.

Power Tools

The carpenter uses many power tools to aid in getting the job done. The quicker the job is done, the more valuable the work of the carpenter becomes. This is called productivity. The more you are able to produce, the more valuable you are. This means the contractor can make money on the job. This means you can have a job the next time there is a need for a good carpenter. Power tools make your work go faster. They also help you to do a job without getting fatigued. Many tools have been designed with you in mind. They are portable and operate from an extension cord.

Table 1-1 *Size of Extension Cords for Portable Tools*

| Cord Length, Feet | Full-Load Rating of the Tool in Amperes at 115 Volts | | | | | |
	0 to 2.0	2.10 to 3.4	3.5 to 5.0	5.1 to 7.0	7.1 to 12.0	12.1 to 16.0
	Wire Size (AWG)					
25	18	18	18	16	14	14
50	18	18	18	16	14	12
75	18	18	16	14	12	10
100	18	16	14	12	10	8
200	16	14	12	10	8	6
300	14	12	10	8	6	4
400	12	10	8	6	4	4
500	12	10	8	6	4	2
600	10	8	6	4	2	2
800	10	8	6	4	2	1
1000	8	6	4	2	1	0

If the voltage is lower than 115 volts at the outlet, have the voltage increased or use a much larger cable than listed.

The extension cord should be the proper size to take the current needed for the tool being used. See Table 1-1. Note how the distance between the outlet and the tool using the power is critical. If the distance is great, then the wire must be larger in size to handle the current without too much loss. The higher the number of the wire, the smaller the diameter of the wire. The larger the size of the wire (diameter), the more current it can handle without dropping the voltage.

Some carpenters run an extension cord from the house next door for power before the building site is furnished power. If the cord is too long or has the wrong size wire, it drops the voltage below 115. This means the saws or other tools using electricity will draw more current and therefore drop the voltage more. Every time the voltage is dropped, the device tries to obtain more current. This becomes a self-defeating phenomenon. You wind up with a saw that has little cutting power. You may have a drill that won't drill into a piece of wood without stalling. Of course the damage done to the electric motor is in some cases irreparable. You may have to buy a new saw or drill. Double-check Table 1-1 for the proper wire size in your extension cord.

Portable saw This is the most often used and abused of carpenter's equipment. The electric portable saw, such as the one shown in Fig. 1-31, is used to cut all 2 × 4s and other dimensional lumber. It is used to cut off rafters. This saw is used to cut sheathing for roofs. It is used for almost every sawing job required in carpentry.

This saw has a guard over the blade. The guard should always be left intact. Do not remove the saw guard. If not held properly against the wood being cut, the saw can kick back and into your leg.

You should always wear safety glasses when using this saw. The sawdust is thrown in a number of direc-

Fig. 1-31 *Portable power saw, the favorite power tool of every carpenter. Note the blade should not extend more than ⅛ inch below the wood being cut. Also note the direction of the blade rotation.*

tions, and one of these is straight up toward your eyes. If you are watching a line where you are cutting, you definitely should have on glasses.

Table saw If the house has been enclosed, it is possible to bring in a table saw to handle the larger cutting jobs. See Fig. 1-32. You can do ripping a little more safely with this type of saw because it has a rip fence. If a push stick is used to push the wood through and past the blade, it is safe to operate. Do not remove the safety guard. This saw can be used for both crosscut and rip. The blade is lowered or raised to the thickness of the wood. It should protrude about ¼ to ½ inch above the wood being cut. This saw usually requires a 1-horsepower motor. This means it will draw about 6.5 amperes to run and over 35 amperes to start. It is best not to run the saw on an extension cord. It should be wired directly to the power source with circuit breakers installed in the line.

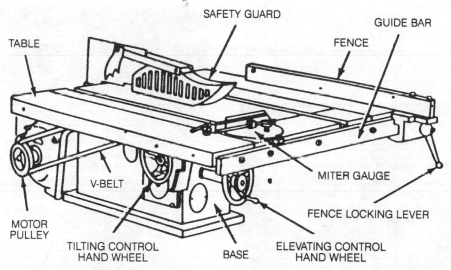

Fig. 1-32 *Table saw.* (Power Tool Division, Rockwell International)

Labels in figure: SAFETY GUARD, GUIDE BAR, TABLE, FENCE, MITER GAUGE, V-BELT, FENCE LOCKING LEVER, MOTOR PULLEY, TILTING CONTROL HAND WHEEL, BASE, ELEVATING CONTROL HAND WHEEL

Radial arm saw This type of saw is brought in only if the house can be locked up at night. The saw is expensive and too heavy to be moved every day. It should have its own circuit. The saw will draw a lot of current when it hits a knot while cutting wood. See Fig. 1-33.

Fig. 1-33 *Radial arm saw.* (DeWalt)

In this model the moving saw blade is pulled toward the operator. In the process of being pulled toward you, the blade rotates so that it forces the wood being cut against the bench stop. Just make sure your left hand is in the proper place when you pull the blade back with your right hand. It takes a lot of care to operate a saw of this type. The saw works well for cutting large-dimensional lumber. It will crosscut or rip. This saw will also do miter cuts at almost any angle. Once you become familiar with it, the saw can be used to bevel crosscut, bevel miter, bevel rip, and even cut cir-

cles. However, it does take practice to develop some degree of skill with this saw.

Router The router has a high-speed type of motor. It will slow down when overloaded. It takes the beginner some time to adjust to *feeding* the router properly. If you feed it too fast, it will stall or burn the edge you're routing. If you feed it too slowly, it may not cut the way you wish. You will have to practice with this tool for some time before you're ready to use it to make furniture. It can be used for routing holes where needed. It can be used to take the edges off laminated plastic on countertops. Use the correct bit, though. This type of tool can be used to the extent of the carpenter's imagination. See Fig. 1-34.

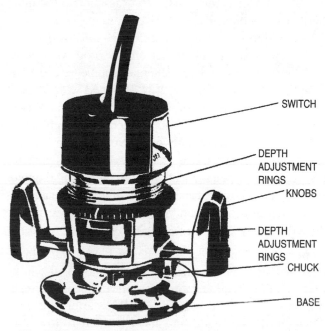

Labels in figure: SWITCH, DEPTH ADJUSTMENT RINGS, KNOBS, DEPTH ADJUSTMENT RINGS, CHUCK, BASE

Fig. 1-34 *The handheld router has many uses in carpentry.*

Saw blades There are a number of saw blades available for the portable, table, or radial saw. They may be standard steel types or they may be carbide-tipped. Carbide-tipped blades tend to last longer. See Fig. 1-35.

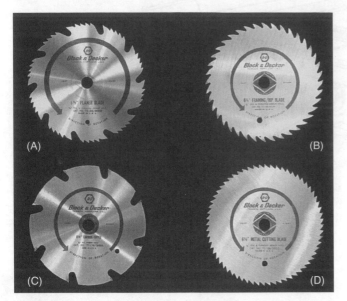

Fig. 1-35 Saw blades. (A) Planer blade, (B) Framing rip blade, (C) Carbide-tipped, (D) Metal cutting blade. (Black & Decker)

Combination blades (those that can be used for both crosscut and rip) with a carbide tip give a smooth finish. They come in 7- to 7¼-inch diameter with 24 teeth. The arbor hole for mounting the blade on the saw is ¾ to ⅝ inch. A safety combination blade is also made in 10-inch-diameter size with 10 teeth and the same arbor hole sizes as the combination carbide-tipped blade.

The planer blade is used to crosscut, rip, or miter hard or soft woods. It is 6½ or 10 inches in diameter with 50 teeth. It too can fit anything from ¾- to ⅝-inch arbors.

If you want a smooth cut on plywood without the splinters that plywood can generate, you better use a carbide-tipped plywood blade. It is equipped with 60 teeth and can be used to cut plywood, Formica, or laminated countertop plastic. It can also be used for straight cutoff work in hardwoods or softwoods. Note the shape of the saw teeth to get some idea as to how each is designed for a specific job. You can identify these after using them for some time. Until you can, mark them with a grease pencil or marking pen when you take them off. A Teflon-coated blade works better when cutting treated lumber.

Saber saw The saber saw has a blade that can be used to cut circles in wood. See Fig. 1-36. It can be used to cut around any circle or curve. If you are making an inside cut, it is best to drill a starter hole first. Then insert the blade into the hole and follow your mark. The saber saw is especially useful in cutting out

Fig. 1-36 Saber saw.

holes for heat ducts in flooring. Another use for this type of saw is to cut holes in roof sheathing for pipes and other protrusions. The saw blade is mounted so that it cuts on the upward stroke. With a fence attached, the saw can also do ripping.

Drill The portable power drill is used by carpenters for many tasks. Holes must be drilled in soleplates for anchor bolts. Using an electric power drill (Fig. 1-37A) is faster and easier than drilling by hand. This drill is capable of drilling almost any size hole through dimensional lumber. A drill bit with a carbide tip enables the carpenter to drill in concrete as well as bricks. Carpenters use this type of masonry hole to insert anchor bolts in concrete that has already hardened. Electrical boxes have to be mounted in drilled holes in the brick and concrete. The job can be made easier and can be more

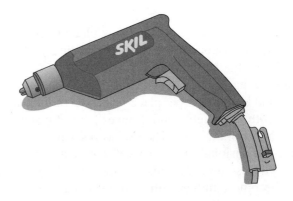

Fig. 1-37A Handheld portable drill.

efficiently accomplished with the portable power hand drill.

The drill has a tough, durable plastic case. Plastic cases are safer when used where there is electrical work in progress.

Carpenters are now using cordless electric drills (Fig. 1-37B). Cordless drills can be moved about the job without the need for extension cords. Improved battery technology has made the cordless drills almost as powerful as regular electric drills. The cordless drill has numbers on the chuck to show the power applied to the shaft. Keep in mind that the higher the number, the greater the torque. At low power settings, the chuck will slip when the set level of power is reached. This allows the user to set the drill to drive screws.

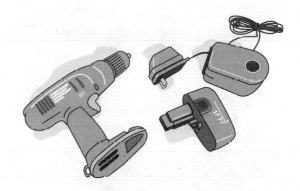

Fig. 1-37D *One charger can be used to charge saw and drill batteries of the same voltage.*

Fig. 1-37B *A cordless hand drill with variable torque.*

Figure 1-37C shows a cordless drill and a cordless saw. This cordless technology is now used by carpenters and do-it-yourselfers. Cordless tools can be obtained in sets that use the same charger system (Fig. 1-37D). An extra set of batteries should be kept charg-

ing at all times and then "swapped out" for the discharged ones. This way no time is lost waiting for the battery to reach full charge. Batteries for cordless tools are rated by battery voltage. High voltage gives more power than low voltage.

As a rule, battery-powered tools do not give the full power of regular tools. However, most jobs don't require full power. Uses for electric drills are limited only by the imagination of the user. The cordless feature is very handy when you are mounting countertops on cabinets. Sanding disks can be placed in the tool and used for finishing wood. Wall and roof parts are often screwed in place rather than nailed. Using the drill with special screwdriver bits can make the job faster than nailing.

Sanders The belt sander shown in Fig. 1-38 and the orbital sanders shown in Fig. 1-39A and B can do almost any required sanding job. The carpenter needs the sander occasionally. It helps align parts properly, especially those that don't fit by just a small amount. The sander can be used to finish off windows, doors, counters, cabinets, and floors. A larger model of the

Fig. 1-37C *A cordless drill and a cordless saw using matching batteries.*

Fig. 1-38 *Belt sander.* (Black & Decker)

(A)

(B)

Fig. 1-39 *Orbital sanders: (A) dual-action and (B) single-action.*
(Black & Decker)

belt sander is used to sand floors before they are sealed and varnished. The orbital or vibrating sanders are used primarily to put a very fine finish on a piece of wood. Sandpaper is attached to the bottom of the sander. The sander is held by hand over the area to be sanded. The operator has to remove the sanding dust occasionally to see how well the job is progressing.

Nailers One of the greatest tools the carpenter has acquired recently is the nailer. See Fig. 1-40. It can drive nails or staples into wood better than a hammer. The nailer is operated by compressed air. The staples and nails are especially designed to be driven by the machine. See Tables 1-2 and 1-3 for the variety of fasteners used with this type of machine. The stapler or nailer can also be used to install siding or trim around a window.

The tool's low air pressure requirements (60 to 90 pounds per square inch) allow it to be moved from place

Fig. 1-40 *Air-powered nailer.* (Duo-Fast)

to place. Nails for this machine (Fig. 1-40) are from 6d to 16d. It is magazine-fed for rapid use. Just pull the trigger.

FOLLOWING CORRECT SEQUENCES

One of the important things a carpenter must do is to follow a sequence. Once you start a job, the sequence has to be followed properly to arrive at a completed house in the least amount of time.

Preparing the Site

Preparing the site may be expensive. There must be a road or street. In most cases the local ordinances require a sewer. In most locations the storm sewer and the sanitary sewer must be in place before building starts. If a sanitary sewer is not available, you should plan for a septic tank for sewage disposal.

Figure 1-41 shows a sewer project in progress. This shows a street being extended. The storm sewer lines are visible, as is the digger. Trees had to be removed first by a bulldozer. Once the sewer lines are in, the roadbed or street must be properly prepared. Figure 1-42 shows the building of a street. Proper drainage is very important. Once the street is in and the curbs are poured, it is time to locate the house.

Figure 1-43 shows how the curb has been broken and the telephone terminal box installed in the weeds. Note the stake with a small piece of cloth on it. This marks the location of the site.

Table 1-2 *Fine Wire Staples for a Pneumatic Staple Driver*

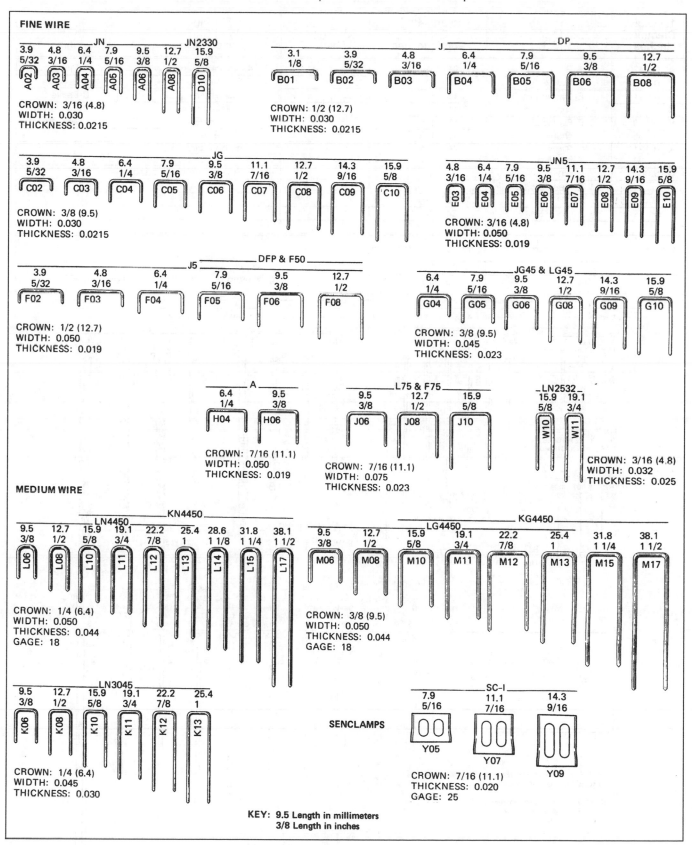

FINE WIRE

| | | | JN | | | | JN2330 |
| 3.9 5/32 A02 | 4.8 3/16 A03 | 6.4 1/4 A04 | 7.9 5/16 A05 | 9.5 3/8 A06 | 12.7 1/2 A08 | | 15.9 5/8 D10 |

CROWN: 3/16 (4.8)
WIDTH: 0.030
THICKNESS: 0.0215

| | | J | | DP | | |
| 3.1 1/8 B01 | 3.9 5/32 B02 | 4.8 3/16 B03 | 6.4 1/4 B04 | 7.9 5/16 B05 | 9.5 3/8 B06 | 12.7 1/2 B08 |

CROWN: 1/2 (12.7)
WIDTH: 0.030
THICKNESS: 0.0215

JG

| 3.9 5/32 C02 | 4.8 3/16 C03 | 6.4 1/4 C04 | 7.9 5/16 C05 | 9.5 3/8 C06 | 11.1 7/16 C07 | 12.7 1/2 C08 | 14.3 9/16 C09 | 15.9 5/8 C10 |

CROWN: 3/8 (9.5)
WIDTH: 0.030
THICKNESS: 0.0215

JN5

| 4.8 3/16 E03 | 6.4 1/4 E04 | 7.9 5/16 E05 | 9.5 3/8 E06 | 11.1 7/16 E07 | 12.7 1/2 E08 | 14.3 9/16 E09 | 15.9 5/8 E10 |

CROWN: 3/16 (4.8)
WIDTH: 0.050
THICKNESS: 0.019

J5 ———— DFP & F50 ————

| 3.9 5/32 F02 | 4.8 3/16 F03 | 6.4 1/4 F04 | 7.9 5/16 F05 | 9.5 3/8 F06 | 12.7 1/2 F08 |

CROWN: 1/2 (12.7)
WIDTH: 0.050
THICKNESS: 0.019

JG45 & LG45

| 6.4 1/4 G04 | 7.9 5/16 G05 | 9.5 3/8 G06 | 12.7 1/2 G08 | 14.3 9/16 G09 | 15.9 5/8 G10 |

CROWN: 3/8 (9.5)
WIDTH: 0.045
THICKNESS: 0.023

A

| 6.4 1/4 H04 | 9.5 3/8 H06 |

CROWN: 7/16 (11.1)
WIDTH: 0.050
THICKNESS: 0.019

L75 & F75

| 9.5 3/8 J06 | 12.7 1/2 J08 | 15.9 5/8 J10 |

CROWN: 7/16 (11.1)
WIDTH: 0.075
THICKNESS: 0.023

LN2532

| 15.9 5/8 W10 | 19.1 3/4 W11 |

CROWN: 3/16 (4.8)
WIDTH: 0.032
THICKNESS: 0.025

MEDIUM WIRE

LN4450 ———— KN4450

| 9.5 3/8 L06 | 12.7 1/2 L08 | 15.9 5/8 L10 | 19.1 3/4 L11 | 22.2 7/8 L12 | 25.4 1 L13 | 28.6 1 1/8 L14 | 31.8 1 1/4 L15 | 38.1 1 1/2 L17 |

CROWN: 1/4 (6.4)
WIDTH: 0.050
THICKNESS: 0.044
GAGE: 18

LG4450 ———— KG4450

| 9.5 3/8 M06 | 12.7 1/2 M08 | 15.9 5/8 M10 | 19.1 3/4 M11 | 22.2 7/8 M12 | 25.4 1 M13 | 31.8 1 1/4 M15 | 38.1 1 1/2 M17 |

CROWN: 3/8 (9.5)
WIDTH: 0.050
THICKNESS: 0.044
GAGE: 18

LN3045

| 9.5 3/8 K06 | 12.7 1/2 K08 | 15.9 5/8 K10 | 19.1 3/4 K11 | 22.2 7/8 K12 | 25.4 1 K13 |

CROWN: 1/4 (6.4)
WIDTH: 0.045
THICKNESS: 0.030

SENCLAMPS

SC-I

| 7.9 5/16 Y05 | 11.1 7/16 Y07 | 14.3 9/16 Y09 |

CROWN: 7/16 (11.1)
THICKNESS: 0.020
GAGE: 25

KEY: **9.5 Length in millimeters**
3/8 Length in inches

Table 1-3 *Seven-Digit Nail Ordering System*

1st Digit: Diameter, Inches	2d Digit: Head	3d and 4th Digits: Length, Inches		5th Digit: Point	6th Digit: Wire Chem. and Finish	7th Digit: Finish
A 0.0475	**A** Brad	**08**	1/2	**A** Diam.-reg.	**A** Std. carbon-galv.	**A** Plain
D 0.072	**C** Flat	**11**	3/4	**E** Chisel	**E** Std. carb. "Weatherex" galv.	**B** Sencote
E 0.0915	**E** Flat/ring shank	**13**	1			**C** Painted
G 0.113		**15**	1 1/4		**G** Stainless steel std. tensile	**D** Painted and sencote
H 0.120	**F** Flat/screw shank	**17**	1 1/2			
J 0.105		**19**	1 3/4		**H** Hardened high-carbon bright basic	
K 0.131	**Y** Slight-headed pin	**20**	1 7/8			
U 0.080		**21**	2		**P** Std. carbon bright basic	
	Z Headless pin	**22**	2 1/8			
		23	2 1/4			
		24	2 3/8			
		25	2 1/2			
		26	2 3/4			
		27	3		EXAMPLE: 10 1/4 ga. (K), flat head (C),	
		28	3 1/4		KC25AAA—2 1/2" (25), regular point (A), std. carb.	
		29	3 1/2		galvanized (A), plain, or uncoated (A) Senco-Nail	

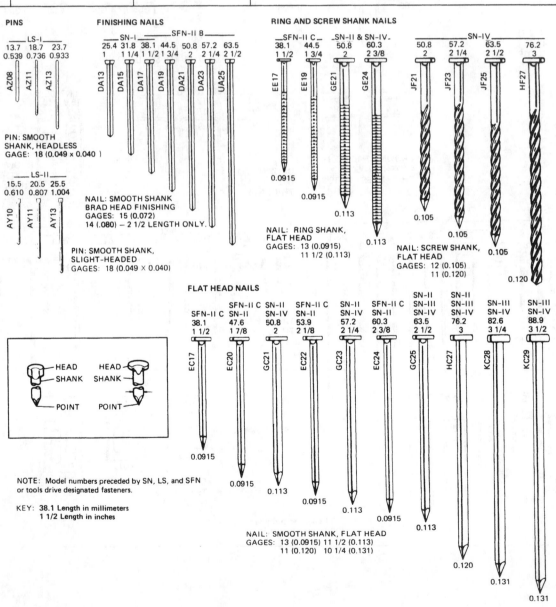

PINS

LS-I
13.7 18.7 23.7
0.539 0.736 0.933
AZ08 AZ11 AZ13

PIN: SMOOTH
SHANK, HEADLESS
GAGE: 18 (0.049 x 0.040)

LS-II
15.5 20.5 25.5
0.610 0.807 1.004
AY10 AY11 AY13

PIN: SMOOTH SHANK,
SLIGHT-HEADED
GAGES: 18 (0.049 × 0.040)

FINISHING NAILS

SN-I   SFN-II B
25.4 31.8 38.1 44.5 50.8 57.2 63.5
1 1 1/4 1 1/2 1 3/4 2 2 1/4 2 1/2
DA13 DA15 DA17 DA19 DA21 DA23 UA25

NAIL: SMOOTH SHANK
BRAD HEAD FINISHING
GAGES: 15 (0.072)
14 (.080) — 2 1/2 LENGTH ONLY.

RING AND SCREW SHANK NAILS

SFN-II C   SN-II & SN-IV
38.1 44.5 50.8 60.3
1 1/2 1 3/4 2 2 3/8
EE17 EE19 GE21 GE24
0.0915
0.0915
0.113
0.113

NAIL: RING SHANK,
FLAT HEAD
GAGES: 13 (0.0915)
11 1/2 (0.113)

SN-IV
50.8 57.2 63.5 76.2
2 2 1/4 2 1/2 3
JF21 JF23 JF25 HF27
0.105
0.105
0.105
0.120

NAIL: SCREW SHANK,
FLAT HEAD
GAGES: 12 (0.105)
11 (0.120)

FLAT HEAD NAILS

SFN-II C SFN-II C SN-II SFN-II C SN-II SFN-II C SN-II SN-II SN-III SN-III
          SN-II   SN-IV SN-II   SN-IV SN-II   SN-III SN-III SN-IV SN-IV
                                                      SN-IV SN-IV
38.1 47.6 50.8 53.9 57.2 60.3 63.5 76.2 82.6 88.9
1 1/2 1 7/8 2 2 1/8 2 1/4 2 3/8 2 1/2 3 3 1/4 3 1/2
EC17 EC20 GC21 EC22 GC23 EC24 GC25 HC27 KC28 KC29
0.0915
0.0915
0.113
0.0915
0.113
0.0915
0.113
0.120
0.131
0.131

HEAD
SHANK
POINT

HEAD
SHANK
POINT

NOTE: Model numbers preceded by SN, LS, and SFN
or tools drive designated fasteners.

KEY: 38.1 Length in millimeters
1 1/2 Length in inches

NAIL: SMOOTH SHANK, FLAT HEAD
GAGES: 13 (0.0915) 11 1/2 (0.113)
11 (0.120) 10 1/4 (0.131)

Fig. 1-41 *Street being extended for a new subdivision.*

Fig. 1-42 *The beginning of a street.*

Fig. 1-43 *Locating a building site and removing the curb for the driveway.*

As you can see in Fig. 1-44, the curb has been removed. A gravel bed has been put down for the driveway.

The sewer manhole sticks up in the driveway. The basement has been dug. Dirt piles around it show how deep the basement really is. However, a closer look shows that the hole isn't too deep. That means the dirt will be pushed back against the basement wall to form a higher level for the house. This will provide drainage

Fig. 1-44 *Dirt from the basement excavation is piled high around a building site.*

away from the house when finished. See Fig. 1-45 for a look at the basement hole.

The Basement

In Fig. 1-46 the columns and the foundation wall have been put up. The basement is prepared in this case with courses of block with brick on the outside. This basement appears to be more of a crawl space under the first floor than a full stand-up basement.

Once the basement is finished and the floor joists have been placed, the flooring is next.

Fig. 1-45 *Hole for a basement.*

Fig. 1-46 *The columns and foundation walls will help support the floor parts.*

The Floor

After the basement or foundation has been laid for the building, the next step is to place the floor over the joists. Note in Fig. 1-47 that the grooved flooring is laid in large sheets. This makes the job go faster and reinforces the floor.

Wall Frames

After the floor is in place and the basement entrance hole has been cut, the floor can be used to support the wall frame. The 2 × 4s or 2 × 6s for the framing can be placed on the flooring and nailed. Once together, they are pushed into the upright position, as in Fig. 1-48. For a two-story house, the second floor is placed on the first-story wall supports. Then the second-floor walls are nailed together and raised into position.

Sheathing

After the sheathing is on and the walls are upright, it is time to concentrate on the roof. See Fig. 1-49. The rafters are cut and placed into position and nailed firmly. See Fig. 1-50. They are reinforced by the proper horizontal bracing. This makes sure they are properly designed for any snow load or other loads that they may experience.

Roofing

The roofing is applied after the siding is on and the rafters are erected. The roofing is completed by applying the proper underlayment and then the shingles. If asphalt shingles are used, the procedure is slightly different from that for wooden shingles. Shingles and roofing are covered in *Miller's Guide to Framing and Roofing*. Figure 1-51 shows the sheathing in place and ready for the roofing.

Fig. 1-47 *Carpenters are laying plywood subflooring with tongue-and-groove joints. This is stronger.* (American Plywood Association)

Fig. 1-49 *Beginning construction of the roof structure.* (Georgia-Pacific)

Fig. 1-48 *Wall frames are erected after the floor frame is built.*

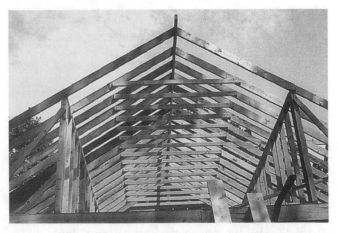

Fig. 1-50 *Framing and supports for rafters.*

Fig. 1-51 *Fiberboard sheathing over the wall frame.*

Fig. 1-53 *Siding applied to building. Note the pattern of the staples.*

Siding

After the roofing, the finishing job will have to be undertaken. The windows and doors are in place. Finish touches are next. The plumbing and drywall may already be in. Then the siding has to be installed. In some cases, of course, it may be brick. This calls for bricklayers to finish up the exterior. Otherwise the carpenter places siding over the walls. Figure 1-52 shows the beginning of the siding at the top left of the picture.

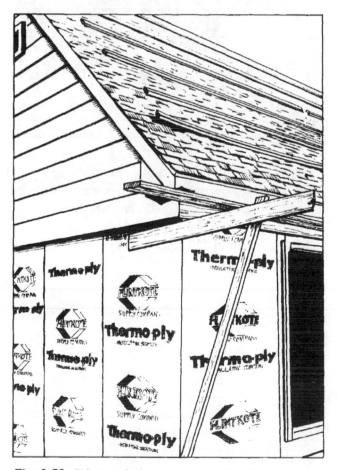

Fig. 1-52 *Siding applied on the top left side of the building.*

Figure 1-53 shows how the siding has been held in place with a stapler. The indentations in the wood show a definite pattern. The siding is nailed to the nail base underneath after a coating of tar paper (felt paper in some parts of the country) is applied to the nail base or sheathing.

Finishing

Exterior finishing requires a bit of caulking with a caulking gun. Caulk is applied to the siding that butts the windows and doors.

Finishing the interior can be done at a more leisurely pace once the exterior is enclosed. The plumbing and electrical work have to be done before the drywall or plaster is applied. Once the wallboard has been finished, the trim can be placed around the edges of the walls, floors, windows, and doors. The flooring can be applied after the finishing of the walls and ceiling. The kitchen cabinets must be installed before the kitchen flooring. There is a definite sequence to all these operations.

As you can imagine, it would be impossible to place roofing on a roof that wasn't there. It takes planning and following a sequence to make sure the roof is there when the roofing crew comes around to nail the shingles in place. The water must be there before you can flush the toilets. The electricity must be hooked up before you can turn on a light. These are reasonable things. All you have to do is sit down and plan the whole operation before starting. Planning is the key to sequencing. Sequencing makes it possible for everyone to be able to do a job at the time assigned to do it.

The Laser Level

The need for plumb walls and level moldings, as well as making various other points straight and level, is paramount in house building. It is difficult in some locations to establish a reference point to check for level windows, doors, and roofs, as well as ceilings and steps.

The laser level (Fig. 1-54) has eliminated much of this trouble in house building. This simple, easy-to-use tool is accurate to within ⅛ inch in 150 feet, and it has become less expensive recently so that even do-it-yourselfers can rent or buy one.

The laser level can generate a vertical reference plane for positioning a wall partition or for setting up forms. See Fig. 1-55. It can produce accurate height gaging and alignment of ceilings, moldings, and horizontal planes and can accurately locate doorways, windows, and thresholds for precision framing and finishing. See Fig. 1-56. The laser level can aid in leveling floors, both indoors and outdoors. It can be used to check stairs, slopes, and drains. The laser beam is easy to use and accurate in locating markings for roof pitches, and it works well in hard-to-reach situations. See Fig. 1-57. The laser beam is generated by two AAA alkaline batteries that will operate for up to 16 hours.

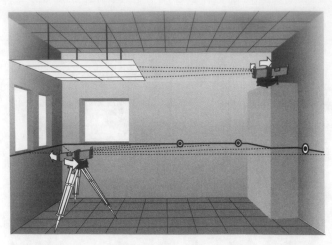

Fig. 1-56 *The laser level can be used for indoor or outdoor leveling of floors, stairs, slopes, drains and ceilings, and moldings around the room.* (Stabila®)

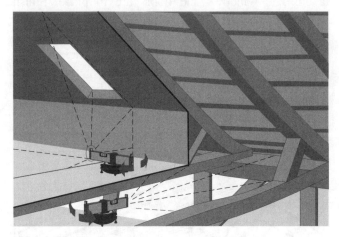

Fig. 1-57 *The laser beam is used to provide easy and accurate location markings on pitches and in hard-to-reach situations.* (Stabilia®)

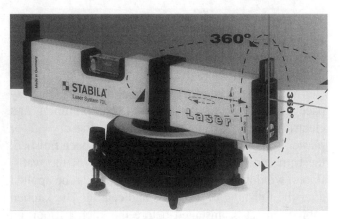

Fig. 1-54 *Laserspirit level moves 360° horizontally and 360° vertically with the optional lens attachment. It sets up quickly and simply with only two knobs to adjust.* (Stabila®)

The combination laser and spirit level quickly and accurately lays out squares and measures plumb. No protective eyewear is needed. The laser operates on a wavelength of 635 nanometers and can have an extended range up to 250 feet.

Fig. 1-55 *The laser level can be used to align ceilings, moldings, and horizontal planes; it produces accurate locations for doorways, windows, and thresholds for precision framing and finishing.* (Stabila®)

2
CHAPTER

Preparing the Site

INITIAL PLANNING IS VITALLY IMPORTANT TO the success of the completed job. Before you begin any construction, you must plan the building on paper. Always keep in mind that it is much cheaper to make a mistake on paper than on site.

In this chapter, you will learn how to develop these skills:

- Locate the boundaries.
- Lay out buildings.
- Use the carpenter's level.
- Prepare for the start of construction.

Each has its importance. Locating the boundaries will allow you to properly lay out the building on the site.

The builder's level will show you how to make sure the building is level. No new leaning Tower of Pisa is needed today. One is enough. It would be very hard to sell one today. That means you should choose the site so that the soil will support the weight of the building.

There is a basic sequence to follow in building. It should be followed for the benefit of those who are supposed to operate as part of the team.

BASIC SEQUENCE

The basic sequence involves the following operations:

1. Cruise the site and plan the job.
2. Locate the boundaries.
3. Locate the building area or areas.
4. Define the site work that is needed.
5. Clear any unwanted trees.
6. Lay out the building.
7. Establish the exact elevations.
8. Excavate the basement or foundation.
9. Provide for access during construction.
10. Start the delivery of materials to the site.
11. Have a crew arrive to start with the footings.

LOCATING THE BUILDING ON THE SITE

The proper location of a building is very important. It would be embarrassing and costly to move a building once it was built. That means a lot of things have to be checked first.

Property Boundaries

First, a clear deed to the land should be established. This can be done in the county courthouse. Check the

records or have someone who is paid for this type of work do it. An abstract of the history of the ownership of the land is usually provided. In Iowa, for instance, the abstract traces ownership back to the Louisiana Purchase of 1803. In New York State the history of the land is traced by owners from the days of the Holland Land Company. Alabama can provide records back to the time the Creek or other Indians owned the land. Each state has its own history and its own procedure for establishing absolute ownership of land. It is best to have proof of this ownership before you start any construction project.

Surveyors should be called in to establish the limits of the property. A plot plan is drawn by the surveyors. This can be used to locate the property. Figure 2-1 is a plot plan showing the location of a house on a lot.

Sidewalks, utilities easements, and other things have to be taken into consideration. The location of the house may be specified by local ordinance. This type of ordinance will usually specify what clearance the house must have on each side. It will probably set the limits of setback from the street. You may also want to plan around trees. Since trees increase the

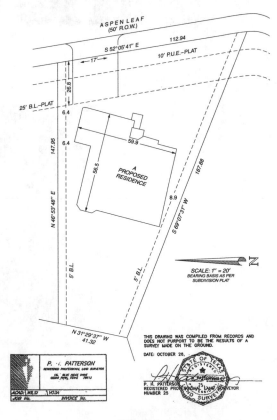

Fig. 2-1 *Plot plan.*

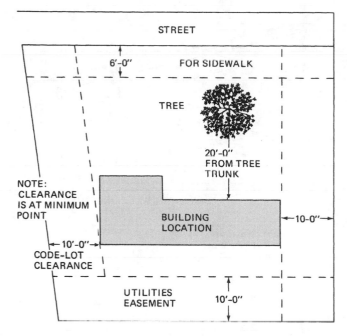

Fig. 2-2 *Site location must be chosen carefully.*

Fig. 2-3 *Rough-cleared lot. Only weeds need to be taken out as the basement is dug.*

value of the property, it is important to save as many as possible. Figure 2-2 shows a sketch of some of these considerations.

An *easement* is the right of the utilities to use the space to furnish electric power, phone service, and gas to your location and to others nearby. This means you have given them permission to string wires or bury lines to provide their services. Keep in mind also the rights of the city or township to supply water and sewers. These may also cut across the property.

Laying Out the Foundation

Layout of the foundation is the critical beginning in house construction. It is a simple but extremely important process. It requires careful work. Make sure the foundation is square and level. You will find all later jobs, from rough carpentry through finish construction and installation of cabinetry, are made much easier.

1. Make sure your proposed house location on the lot complies with local regulations.

2. Set the house location, based on required setbacks and other factors, such as the natural drainage pattern of the lot. Level or at least rough-clear the site. See Fig. 2-3.

3. Lay out the foundation lines. Figure 2-4 shows the simplest method for locating these. Locate each outside corner of the house and drive small stakes into the ground. Drive tacks into the tops of the stakes. This is to indicate the outside line of the foundation wall. This is not the footings limit but

the outside wall limit. Next check the squareness of the house by measuring the diagonals, corner to corner, to see that they are equal. If the structure is rectangular, all diagonal measurements will be equal. You can check squareness of any corner by measuring 6 feet down one side, then 8 feet down the other side. The diagonal line between these two end points should measure exactly 10 feet. If it doesn't, the corner isn't truly square. See Fig. 2-5.

4. After the corners are located and squared, drive three 2×4 stakes at each corner as shown in Fig. 2-4. Locate these stakes 3 feet and 4 feet outside the actual foundation line. Then nail 1×6 batter boards horizontally so that their top edges are all level and at the same grade. Levelness will be checked later. Hold a string line across the tops of opposite batter boards at two corners. Using a plumb bob, adjust the line so that it is exactly over the tacks in the two corner stakes. Cut saw kerfs ¼ inch deep where the line touches the batter boards so that the string lines may be easily replaced if they are broken or disturbed. Figure 2-6 shows how carpenters in some parts of the country use a nail instead of the saw kerf to hold the thread or string. Figure 2-7 shows how the details of the location of the stake are worked out. This one is a 3 – 4 – 5 triangle, or 9 feet and 12 feet on the sides and 15 feet on the diagonal. If you use 6, 8, and 10 feet, you get a 3 – 4 – 5 triangle also. This means $6 \div 2 = 3$, $8 \div 2 = 4$, and $10 \div 2 = 5$. In the other example, 9 feet $\div 3 = 3$, 12 feet $\div 3 = 4$, and 15 feet $\div 3 = 5$. So you have a 3 – 4 – 5 triangle in either measurement. Other combinations can be used, but these are the most common. Cut all saw kerfs the same depth. This is done so that the

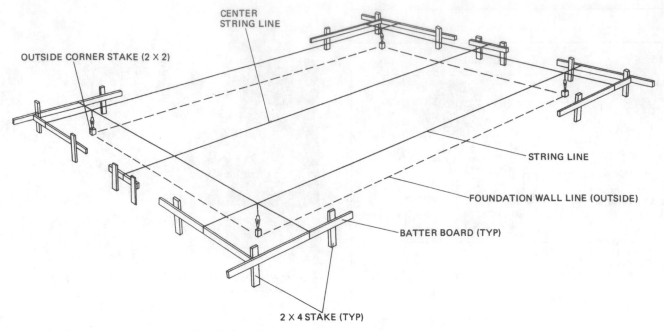

Fig. 2-4 *Staking out a basement. (American Plywood Association)*

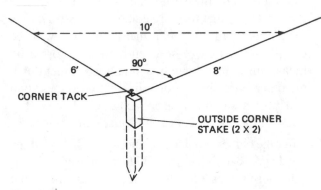

Fig. 2-5 *Squaring the corner and marking the point. (American Plywood Association)*

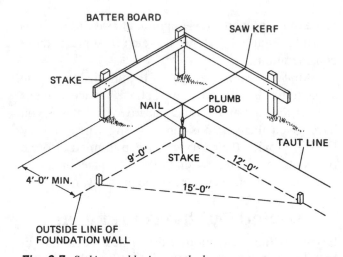

Fig. 2-7 *Staking and laying out the house. (Forest Products Laboratory)*

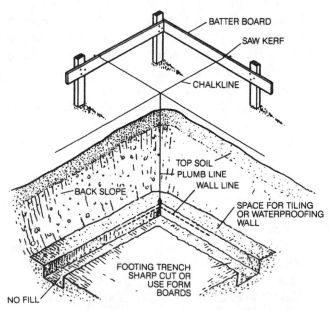

Fig. 2-6 *Note the location of the nails on the batter board. (U.S. Department of Agriculture)*

string line not only defines the outside edges of the foundation but also will provide a reference line. This ensures uniform depth of footing excavation. When you have made similar cuts in all eight batter boards and strung the four lines in position, the outside foundation lines are accurately established.

5. Next, establish the lengthwise girder location. This is usually on the centerline of the house. Double-check your house plans for the exact position. This is done because occasionally the girder will be slightly off the centerline to support an interior bearing wall. To find the line, measure the correct distance from the corners. Then install batter boards and locate the string line as before.

6. Check the foundation for levelness. Remember that the top of the foundation must be level around the entire perimeter of the house. The most accurate and simplest way to check this is to use a surveyor's level. This tool will be explained later in this chapter. The next best approach is to ensure that batter boards, and thus the string lines, are all absolutely level. You can accomplish this with a 10- to 14-foot-long piece of straight lumber. See Fig. 2-8A. Judge the straightness of the piece of lumber by sighting along the surface. Use this straightedge in conjunction with a carpenter's level. Laser levels (Fig. 2-8B) can also be used instead of a long board. Make sure the laser is level and at the right height. Then use the red dot to indicate the height of each leveling stake. Next drive temporary stakes around the house perimeter. The distance between them should not exceed the length of the straightedge. Then place one end of the straightedge on a batter board. Check for exact levelness. See Fig. 2-8. Drive another stake to the same height. Each time a stake is driven, the straightedge and level should be reversed end for end. This should ensure close accuracy in establishing the height of each stake with reference to the batter board. The final check on overall levelness comes when you level the last stake with the batter board where you began. If the straightedge is level here, then you have a level foundation baseline. During foundation excavation, the corner stakes and temporary leveling stakes will be removed. This stresses the importance of the level

batter boards and string line. The corners and foundation levelness must be located using the string line.

THE BUILDER'S LEVEL

Practically all optical sighting and measuring instruments can be termed *surveying instruments*. Surveying, in its simplest form, simply means accurate measuring. Accurate measurements have been a construction requirement ever since humans started building things.

How Does It Work?

Even during the days of pyramid building, humans recognized the fact that the most accurate distant measurements were obtained with a perfectly straight line of sight. The basic principle of operation for today's modern instruments is still the same. A line of sight is a perfectly straight line. The line does not dip, sag, or curve. It is a line without weight and is continuous.

Any point along a level line of sight is exactly level with any other point along that line. The instrument itself is merely the device used to obtain this perfectly level line of sight for measurements.

Three Main Parts of a Builder's Level

1. *The telescope* (Fig. 2-9). The telescope is a precision-made optical sighting device. It has a set of carefully ground and polished lenses. They produce a clear, sharp, magnified image. The magnification of a telescope is described as its power. An 18-power

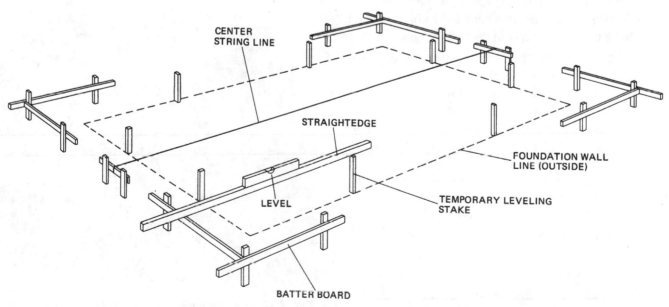

CENTER
STRING LINE

STRAIGHTEDGE

FOUNDATION WALL
LINE (OUTSIDE)

LEVEL

TEMPORARY LEVELING
STAKE

BATTER BOARD

Fig. 2-8A *Leveling the batter boards.* (American Plywood Association)

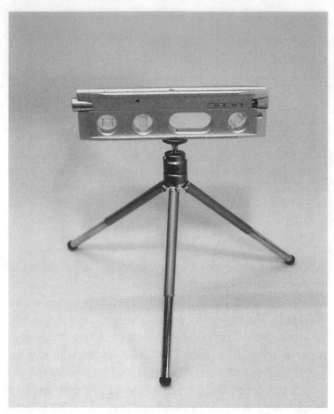

Fig. 2-8B *A laser level on a short tripod can be used on batter boards and leveling stakes.*

Fig. 2-9 *The telescope on an optical level.* (David White Instruments)

telescope will make a distant object appear 18 times closer than when it is viewed with the naked eye. Crosshairs in the telescope permit the object sighted on to be centered exactly in the field of view.

2. *The leveling vial* (Fig. 2-10). Also called the *bubble,* the leveling vial works just like the familiar carpenter's level. However, it is much more sensitive and accurate in this instrument. Four leveling screws on the instrument base permit the user to center (level) the vial bubble perfectly and thus establish a level line of sight through the telescope. A vital first step in instrument use is leveling. Instrument vials are available in various degrees of sensitivity. In general, the more sensitive the vial, the more precise the results that may be obtained.

Fig. 2-10 *The leveling vial on an optical level.* (David White Instruments)

3. *The circle* (Fig. 2-11). The perfectly flat plate upon which the telescope rests is called the circle. It is marked in degrees and can be rotated in any horizontal direction. With the use of an index pointer, any horizontal angle can be measured quickly. Most instruments have a *vernier scale.* An additional scale is subdivided. It divides degrees into minutes. There are 60 minutes in each degree. There are 360° in a circle.

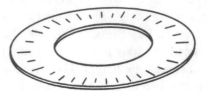

Fig. 2-11 *The circle on an optical level.* (David White Instruments)

Preparing the Instrument

Figure 2-12 shows a builder's level on site. Leveling the instrument is the most important operation in preparing the instrument for use.

Leveling the instrument First, secure the instrument to its tripod and proceed to level it as follows. Figure 2-13A shows the type of tripod used to support the instrument. The target pole is shown in Fig. 2-13B.

Place the telescope directly over one pair of opposite leveling screws. (See Fig. 2-14.) Turn the screws

Fig. 2-12 *Using the optical (or builder's) level on the job.* (David White Instruments)

(A)

(B)

Fig. 2-13 *(A) The tripod for an optical level, (B) The rod holder for use with an optical level. (David White)*

TURNING BOTH LEVELING SCREWS "IN" MOVES BUBBLE TO RIGHT

TURNING BOTH LEVELING SCREWS "OUT" MOVES BUBBLE TO LEFT

SIDE VIEW

TOP VIEW

LEVEL VIAL WITH BUBBLE CENTERED

Fig. 2-15 *Adjusting the leveling screws and watching the bubbles for level. (David White Instruments)*

directly under the scope in opposite directions at the same time (see step 5 in Fig. 2-14) until the level-vial bubble is centered. The telescope is then given a quarter (90°) turn. Place it directly over the other pair of leveling screws (step 2 of Fig. 2-14). The leveling operation is then repeated. Then recheck the other positions (steps 3 and 4 of Fig. 2-14) and make adjustments. This may not be necessary. Adjust if necessary. When leveling is completed, it should be possible to turn the telescope in a complete circle without any changes in the position of the bubble. See Fig. 2-15.

With the instrument leveled, you know that, since the line of sight is perfectly straight, any point on the line of sight will be exactly level with any other point. The drawing in Fig. 2-16 shows how exactly you can check the difference in height (elevation) between two points. If the rod reading at *B* is 3 feet and the reading at *C* is 4 feet, you know that point *B* is 1 foot higher than point *C*. Use the same principle to check if a row of windows is straight or if a foundation is level. Or you can check how much a driveway slopes.

Staking out a house Start at a previously chosen corner to stake out the house. Sight along line *AB* of Fig. 2-17 to establish the front of the house. Measure the desired distance to *B* and mark it with a stake.

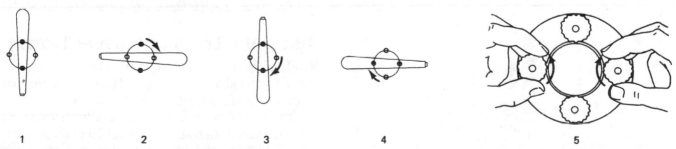

1   2   3   4   5

Fig. 2-14 *Adjusting the screws on the level-transit will level it. Note how it is leveled with two screws, then moved 90° and leveled again. (David White Instruments)*

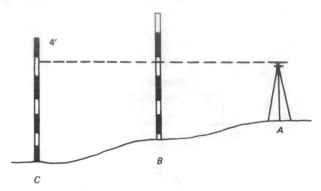

Fig. 2-16 *Finding the elevation with a level.* (David White Instruments)

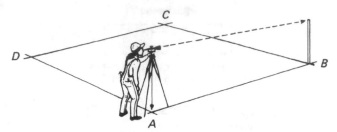

Fig. 2-17 *How to stake out a house on a building lot using a builder's level.* (David White Instruments)

Swing the telescope 90° by the circle scale. Mark the desired distance to *D*. This gives you the first corner. All the others are squared off in the same manner. You're sure all foundation corners are square, and all it took was a few minutes of setup time. See Fig. 2-18.

This method eliminates the use of the old-fashioned string line–tape–plumb bob methods.

The Level-Transit

There are two types of levels used for building sites. The level and the level-transit are the two instruments used. The level has the telescope in a fixed horizontal position but can move sideways 360° to measure horizontal angles. It is usually all that is needed at a building site for a house. See Fig. 2-19.

A combination instrument is called a level-transit. The telescope can move in two directions. It can move up and down 45° as well as from side to side 360°. See Fig. 2-20. It can measure vertical as well as horizontal angles.

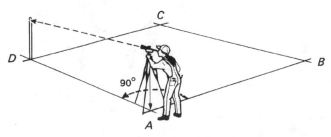

Fig. 2-18 *Squaring the other corner in laying out a building on a lot.* (David White Instruments)

Fig. 2-19 *Builder's level.* (David White Instruments)

Fig. 2-20 *Level-transit.* (David White Instruments)

A lock lever or levers permit the telescope to be securely locked in a true level position for use as a level. A full transit instrument, in addition to the features just mentioned, has a telescope that can rotate 360° vertically.

The level-transit is shown in operation in Fig. 2-21.

Using the Level and Level-Transit

Reading the circle and vernier The 360° circle is divided in quadrants (0 to 90°). The circle is marked by degrees and numbered every 10°. See Fig. 2-22.

To obtain degree readings, it is only necessary to read the exact degree at the intersection of the zero index mark on the vernier and the degree mark on the circle (or on the vertical arc of the level-transit).

Fig. 2-21 *Using the level-transit on the site.* (David White Instruments)

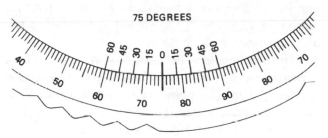

Fig. 2-22 *Reading the circle.* (David White Instruments)

For more precise readings, the vernier scale is used. See Fig. 2-23. The vernier lets you subdivide each whole degree on the circle into fractions, or minutes. There are 60 minutes in a degree. If the vernier zero does not line up exactly with a degree mark on the circle, note the last degree mark passed and, reading up the vernier scale, locate a vernier mark that coincides with a circle mark. This will indicate your reading in degrees and minutes.

Hanging the plumb bob To hang the plumb bob, attach a cord to the plumb bob hook on the tripod. Knot the cord as shown in Fig. 2-24.

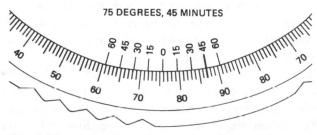

Fig. 2-23 *Reading the circle and vernier.* (David White Instruments)

Fig. 2-24 *To hang a plumb bob, attach a cord to the plumb bob hook on the tripod and knot the cord as shown here.* (David White Instruments)

If you are setting up over a point, attach the plumb bob. Move the tripod and instrument over the approximate point. Be sure the tripod is set up firmly again. Shift the instrument on the tripod head until the plumb bob is directly over the point. Then set the instrument leveling screws again to level the instrument.

Power The power of a telescope is rated in terms of magnification. It may be $24\times$ or $37\times$. The $24\times$ means the telescope is presenting a view 24 times as close as you could see it with the naked eye. Some instruments are equipped with a feature that lets you zoom in from $24\times$ to $37\times$. It increases the effective reading range of the instrument more than 42 percent. It also permits greater flexibility in matching range, image, and light conditions. Use low power for brighter images in dim light. Since it gives a wider field of view, it is also handy in locating targets. Low power also provides better visibility for sighting through heat waves. See Fig. 2-25.

High power is used for sighting under bright light conditions. It is used for long-range sighting and for more precise rod readings.

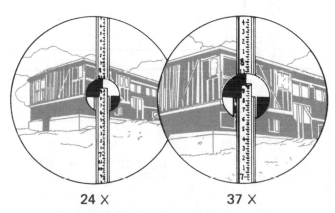

24 X   37 X

Fig. 2-25 *Variable instrument power is available.* (David White Instruments)

Rod Leveling rods are a necessary part of the transit leveling equipment. Rods are direct-reading with large gradations. All rods are equipped with a tough, permanent polyester film scale that will not shrink or expand. This is important when you consider the gradations can be 1/100 of a foot. Figure 2-26 shows a leveling rod with the target attached at 4 feet 5¼ inches. The target (Fig. 2-27) can be moved by releasing a small clamp in the back. Figure 2-28 shows a tape and the gradations. They are ⅛ inch wide and ⅛ inch apart. The tape is marked in feet, inches, and eighths of an inch. Feet are numbered in red. A three-section rod extends to 12 feet. A two-section rod extends to 8 feet 2 inches.

The rod holder is directed by hand signals from the surveyor behind the transit. The hand signals are easy to understand, since they motion in the direction of desired movement of the rod.

Fig. 2-28 *Tape face on the rod. This one is marked in feet, inches, and eighths of an inch.* (David White Instruments)

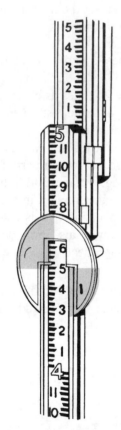

Fig. 2-26 *Leveling rod made of wood.* (David White Instruments)

Fig. 2-27 *Target that fits wood rods.* (David White Instruments)

Establishing Elevations

Not all lots are flat. That means there is some kind of slope to be considered when you are digging the basement or locating the house. The level can help establish what these elevation changes are. From the grade line you establish how much soil will have to be removed for a basement. The grade line will also determine the location of the floor.

The benchmark is the place to start. A benchmark is established by surveyors when they open a section to development. This point is a reference to which the lot you are using is tied. The lot is so many feet in a certain direction from a given benchmark.

The benchmark may appear as a mark or point on the foundation of a nearby building. Sometimes it is the nearby sidewalk, street, or curb that is used as the level reference point.

The grade line is established by the person who designed the building. This line must be accurately established. Many measurements are made from this line. It determines the amount of earth removed from the basement or for the foundation footings.

Using the Leveling Rod

Use a leveling rod and set it at any point where you want to check the elevation. Sight through the level or transit-level to the leveling rod. Take a reading by using the crosshair in the telescope. Move the rod to another point that is to be established. Now raise or lower the rod until the reading is the same as for the first point. This means the bottom of the rod is at the same elevation as the original point.

One person will hold the rod level. Another will move the target up or down until the crosshair in the telescope comes in alignment with that on the target. The difference between the two readings tells you what the elevation is.

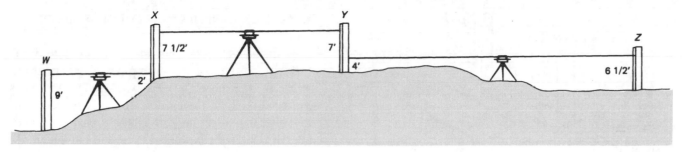

Fig. 2-29 *Getting the elevation when the two points cannot be viewed from a single point.*

Figure 2-29 shows how the difference in elevation between two points that are not visible from a single point is determined.

If point Z cannot be seen from point W, then you have to set the transit up again at two other points, such as X and Y. Take the readings at each location; then you will be able to determine how much of the soil has to be removed for a basement.

PREPARING THE SITE
Clearing

One of the first things to do in preparing the site for construction is to clear the area where the building will be located. Look over the site. Determine if there are trees in the immediate area of the house. If so, mark the trees to be removed. This can be done with a spray can of paint. Put an X on those to be removed or a line around them. In some cases, people have marked those that must go with a piece of cloth tied to a limb.

Also, make sure those that are staying are not damaged when the heavy equipment is brought onto the site. Scarring of trees can cause them to die later. Covering them more than 12 inches will probably kill them also. You have to cut off a part of the treetop. This helps it survive the covering of the roots.

Don't dig the sewer trench or the water lines through the root system of the trees to be saved. This can cause the tops of the trees to die later, and in some cases it will kill them altogether.

Make a rough drawing of the location of the house and the trees to be saved. Make sure the persons operating the bulldozer and digger are made aware of the effort to save trees.

Cutting trees Keep in mind that removing trees can also be profitable. You can cut the trees into small logs for use in fireplaces. This has become an interest of many energy conservationists. The brush and undergrowth can be removed with a bulldozer or other type of equipment. Do not burn the brush or the limbs without checking with local authorities. There is always someone who is interested in hauling off the accumulation of wood.

Stump removal In some cases a tree stump is left and must be removed. There are a number of ways of doing this. One is to use a winch and pull it up by hooking the winch to some type of power takeoff on a truck, tractor, or heavy equipment. You could dig it out, but this can take time and may require too much effort in most cases.

The use of explosives to remove the stump is not permitted in some locations. Better check with the local police before you set off the blast.

The best way is to use the bulldozer to uproot the entire stump or tree. It all depends upon the size of the tree and the size of the equipment available for the job. Anyway, be sure the lot is cleared so the digging of the basement or footings can take place.

Excavation

A house built on a slab does not require any extensive excavation. One-piece or monolithic slabs are used on level ground and in warm climates. In cold climates, where the frostline penetrates deeper, or in areas where drainage is a problem, a two-piece slab has to be used. Figures 2-30 and 2-31 both show two-piece slab foundations.

Slab footings must rest beneath the frostline. This gives stability in the soil. The amount of reinforcement needed for a slab varies. The condition of the soil and the weights to be carried determine the reinforcement. Larger slabs and those on less stable soil need more reinforcement.

The top of the slab must be 8 inches above ground. This allows moisture under the slab to drain away from the building. It also gives you a good chance to spot termites building their tunnels from the earth to the floor of the house. The slab should always rest slightly above the existing grade. This is done to provide for runoff water during a rainstorm.

Basements A basement is the area usually located underground. It provides most homes with a lot of

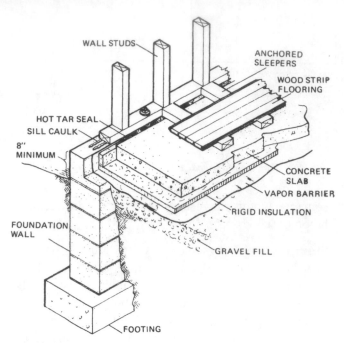

Fig. 2-30 *Two-piece slab with block foundation wall.* (Forest Products Laboratory)

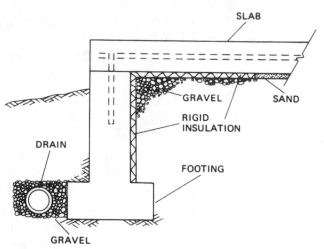

Fig. 2-31 *Two-piece slab with poured wall and footer.*

storage. In some, it is a place to do the laundry. It also serves as a place to locate the heating and cooling units. If a basement is desired, it must be dug before the house is started. The footings must be properly poured and seasoned. Seasoning should be done before poured concrete or concrete block is used for the wall. Some areas now use treated wood walls for a basement.

Figure 2-32 shows a basement dug for use in colder climates. Trenches from the street to the basement must be provided for the plumbing and water. Utilities may be buried also. If they are, the electric, phone, and gas lines must also be located in trenches or buried after the house is finished. It is a good idea to notify the utility companies so that they can schedule the installation of their services when you are ready for them.

Some shovel work may have to be done to dig the basement trenches for the sewer pipes. This is done after the basement has been leveled by machines. As you can see from Fig. 2-32, the basement may also need shovel work after the digger has left. Note the cave-ins and dirt slides evident in the basement excavation in Fig. 2-32.

The basement has to be filled later. Gravel is used to form a base for the poured concrete floor.

The high spots in the basement must be removed by shovel. Proper-size gravel should be spread after the sewer trench is filled. You may have to tamp the gravel to make sure it is properly level and settled. Do this before the concrete mixer is called for the floor job.

The footings have to be poured first. They are boxed in and poured before anything is done in the way of the basement floor. In some instances drain tile must be installed inside or outside the footing. The tile is allowed to drain into a sump. In other locations no drainage is necessary because of soil conditions.

PROVIDING ACCESS DURING CONSTRUCTION

The first thing to be established is who is to be on the premises. Check with your insurance company about liability insurance. This is done in case someone is hurt on the location. Also decide who should be kept out. You also have to decide how access control is to be set up. It may be done with a fence or by an alert guard or dog. These things do have to be considered before the construction gets underway. If equipment is left at the site, who is responsible? Who will pay for vandalism? Who will repair damage caused by wind, hail, rain, lightning, or tornado?

Materials Storage

Where will materials for the job be stored? In Fig. 2-33 you can see how plywood is stored. What happens if someone decides to haul off some of the plywood? Who is responsible? What control do you have over the stored materials after dark?

Figure 2-34 shows plywood bundles broken open. This makes it easy for single sheets to disappear. With the current price of plywood, it becomes important to plan some type of storage facility on the site.

Some of the shingles in Fig. 2-35 may be hard to find if the wind gets to the broken bundle. What's to stop children playing on the site during off hours? They can also take the shingles and spread them over the landscape.

Fig. 2-32 *Excavation for a basement.*

Fig. 2-33 *Storing plywood on the site.*

Fig. 2-34 *Broken bundle of plywood sheathing.*

Storage of bricks can be a problem. See Fig. 2-36. They are expensive and can easily be removed by someone with a small truck. It is very important to have some type of on-site storage. It is also very important to make sure that materials are not delivered before they are needed. Some type of materials inventory has to be maintained. This may be worth a person's time. The location of the site is a major factor in the disappearance of materials. Location has a lot to do with the liability coverage needed from insurance companies.

Temporary buildings Some building sites have temporary structures to use as storage. In some cases the plans for the building are also stored in the toolshed. Covered storage is used in some locations where rain and snow can cause a delay by wetting the lumber, sand, or cement. If you are using drywall, you will need to keep it dry. In most instances it is not delivered to the site until the house is enclosed.

Some construction sheds are made on the scene. In other cases the construction shed may be delivered to the site on a truck. It is picked up and moved away once the building can be locked.

Mobile homes have been used as offices for supervisors. This usually is the case when a number of houses are made by one contractor and all are located in one row or subdivision.

A garage can be used as the headquarters for the construction. The garage is enclosed. It is closed off by

Fig. 2-35 *Broken bundle of shingles.*

Fig. 2-36 *Storing bricks on site.*

doors, so it can be locked at night. Since the garage is easy to close off, it becomes the logical place to take care of paperwork. It also becomes a place to store materials that should not get wet.

When building a smaller house, the carpenter takes everything home at the end of the working day. The carpenter's car or truck becomes the working office away from home. Materials are scheduled for delivery only when actually needed. In larger projects some local office is needed, so the garage, toolshed, mobile home, or construction shack is used.

Storing construction materials Storing construction materials can be a problem. It requires a great deal of effort to make sure the materials are on the job when needed. If delivery schedules are delayed, work has to stop. This puts people out of work.

If materials are stored on the job, make sure they are neatly arranged. This prevents accidents such as tripping over scattered materials. Sand should be delivered and placed out of the way. Keep it out of the normal traffic flow from the street to the building.

Everything should be kept in some order. This means you know where things are when you need them. Then you don't have to plow through a mound of

supplies just to find a box of nails. Everything should be laid out according to its intended order of use.

Lumber should be kept flat. This prevents warpage, cupping, and twisting. Plywood should be protected from rain and snow if it is interior grade. In any case it should not be allowed to become soaked. Keep it flat and covered.

Humidity control is important inside a house. This is especially true when you're working with drywall. It should be allowed to dry by keeping the windows open. Too much humidity can cause the wood to twist or warp.

Temporary Utilities

You will need electricity to operate the power tools. Power can be obtained by using a long extension cord from the house nearby. Or, you may have to arrange for the power company to extend a line to the side and put in a meter on a pole nearby.

Water is needed to mix mortar. The local line will have to be tapped, or you may have to dig the well before you start construction. It all depends upon where the building site is located. If the house is being built near another, you may want to arrange with the neighbors to supply water with a hose to their outside faucet. Make sure you arrange to pay for the service.

Waste Disposal

Every building site has waste. It may be human waste or paper and building-material wrappings. Human waste can be controlled by renting a Porta John or a Johnny-on-the-Spot. This can prevent the house from smelling like a urinal when you enter. The sump basin should not be used as a urinal. It does leave an odor to the place. Besides, it is unsanitary.

Wastepaper can be burned in some localities. In others burning is strictly forbidden. You should check before you arrange to have a large bonfire for getting rid of the trash, cut lumber ends, paper, and loose shingles. There are companies that provide a trash-collecting service for construction areas. They leave the place *broom clean*. It leaves a better impression of the contractor when a building is delivered in order, without trash and wood pieces lying around. If you go to the trouble of building a fine home for someone, the least you can do is to deliver it in a clean condition. After all, this is going to be a home.

Arranging Delivery Routes

Damage to the construction site by delivery trucks can cause problems later. You should arrange a driveway by putting in gravel at the planned location of the drive.

Get permission and remove the curb at the entrance to the driveway. Make sure deliveries are made by this route. Pile the materials so that they are arranged in an orderly manner and can be reached when needed.

Concrete has to be delivered to the site for the basement, foundation, and garage floor. Be sure to allow room for the ready-mix truck to get to these locations. Lumber is usually strapped together. Make allowance for bundles of lumber to be dumped near the location where they will be needed.

Make sure the nearby plants or trees are protected. This may require a fence or stakes. Some method should be devised to keep the trucks, diggers, and earthmovers from destroying natural vegetation.

Access to the building site is important. If this is the first house in the subdivision, or if it is located off the road, you have to provide for delivery of materials. You may have to put in a temporary road. This should be a road that can be traveled in wet weather

without the delivery trucks becoming bogged down or stuck.

As you can see, it takes much planning to accomplish a building program that will come off smoothly. The more planning you do ahead of time, the less time will be spent trying to obtain the correct permissions and deliveries.

The key to a successful building program is planning. Make a checklist of the items that need attention beforehand. Use this checklist to keep yourself current with the delivery of materials and permissions.

It is assumed here the proper financial arrangements have been made before construction begins.

Check the floor plan of the proposed house before proceeding. Make sure everything desired is included. Later changes can get very expensive. See Fig. 2-37.

Also check the sketch of the proposed house to make sure the exterior is what you want. See Fig. 2.38.

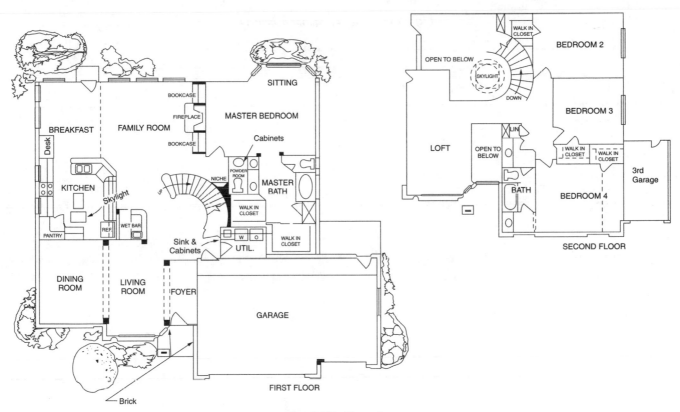

Fig. 2-37 *Floor plan.*

Fig. 2-38 *Sketch of exterior of house.*

3
CHAPTER

Laying Footings & Foundations

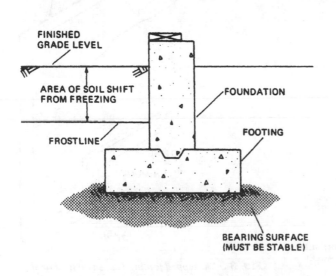

FINISHED GRADE LEVEL

AREA OF SOIL SHIFT FROM FREEZING

FOUNDATION

FOOTING

FROSTLINE

BEARING SURFACE (MUST BE STABLE)

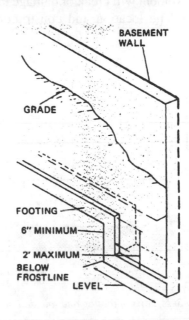

BASEMENT WALL

GRADE

FOOTING

6" MINIMUM

2' MAXIMUM BELOW FROSTLINE

LEVEL

PEOPLE OFTEN THINK THAT FOOTINGS AND foundations are the same. Actually, the footing is the lowest part of the building and carries the weight. The foundation is the wall between the footing and the rest of the building. In this chapter you will learn how to

- Design footings and foundations
- Locate corners and lines for forms
- Check the level of footing and foundation excavation
- Make the forms for footings
- Make the forms for foundations
- Reinforce the forms as required
- Mix or select concrete for usage
- Pour the concrete into the forms
- Finish concrete in the forms
- Embed anchor systems in forms
- Waterproof foundation walls if needed
- Make necessary drainage systems

INTRODUCTION

Footings bear the weight of the building. They spread the weight evenly over a wide surface. Figure 3-1 shows the three parts of a footing system. These parts are the bearing surface, the footing, and the foundation. The bearing surface must be located beneath the frostline on firm and solid ground. The frostline is the deepest level at which the ground will freeze in the wintertime. Moisture in the ground above the frostline will freeze and thaw. When it does, the ground moves and shifts. The movement will break or damage the footing or foundation. The location and construction of the

footing are very important. Think of the weight of all the lumber, concrete, stone, and furniture that must be supported by this layer. All these must be supported without sinking or moving.

Footings may be made in several ways. There are flat footings, stepped footings, pillared footings, and pile footings. The flat footing, as in Fig. 3-2, is the easiest and simplest footing to make because it is all on one level. The stepped footing is used on sides of hills as in Fig. 3-3. The stepped footing is like a series of short flat footings at different levels, much like a flight of steps. By making this type of footing, no special digging (excavation) is needed. The third footing type is the pillared footing. See Fig. 3-4. The pillared foot-

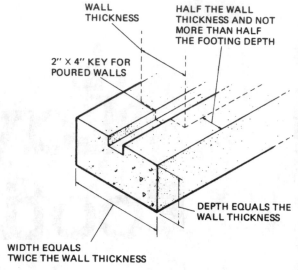

Fig. 3-2 *Regular flat footing.* (Forest Products Laboratory)

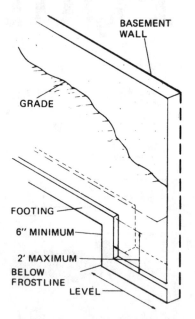

Fig. 3-3 *A stepped footing is used on hills or slopes.* (Forest Products Laboratory)

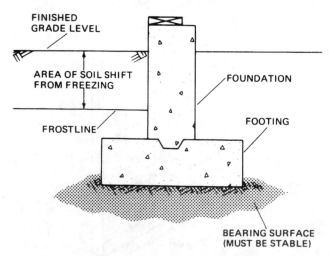

Fig. 3-1 *Parts of footing and foundation.*

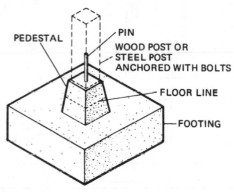

Fig. 3-4 *The pillar or post footing may be square or round.* (Forest Products Laboratory)

ing is used in many locations where the soil is evenly packed and little settling occurs. It consists of a series of pads, or feet. Columns are then built on the pads and the building rests upon the columns. Buildings with either flat or stepped footings usually have pillared footings in the center. This is so because buildings are too wide to support the full weight without support in the middle areas.

Pile footings are the fourth type and are used where soil is loose, unstable, or very wet. As in Fig. 3-5, long columns are put into the ground. These are long enough to reach solid soil. The columns may be made of treated wood or concrete. The wooden piles are driven into the ground by pounding. The concrete piles are made by drilling a hole and filling it with concrete. Pads or caps are then put over the tops of the piles.

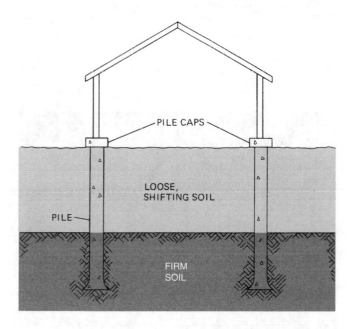

Fig. 3-5 *Pile footings reach through water or shifting soils.*

SEQUENCE

No matter what type of footing and foundation is used, a certain sequence should be followed. The sequence can change slightly according to the method involved. For example, the footing and foundation can be poured in one solid concrete piece. However, many footings are made separately and the concrete foundations are built on top of the footings. In both cases, the sequence is similar. The basic sequence is as follows:

1. Find the amount of site preparation needed.
2. Lay out footing and foundation shape.
3. Excavate to proper depth.
4. Level the footing corners.
5. Build the footing forms.
6. Reinforce the forms as needed.
7. Estimate concrete needs.
8. Pour the concrete footing.
9. Build the foundation forms.
10. Reinforce the forms as needed.
11. Pour the concrete into forms.
12. Finish the concrete and embed anchors.
13. Remove the forms.
14. Waterproof and drain as required.

LAYING OUT THE FOOTINGS

Footings are the bottom of the building and must hold up the weight of the building. Two factors are involved in finding the correct shape and size. The first is the strength or solidness of the soil. The second factor is the width and depth of the footing for the weight of the building in that type of soil.

Soil Strength

Soil strength refers to how dense and solid the soil is packed. It also refers to how stable or unmoving the soil is. Some soils are very hard only when dry. Others keep the same strength whether they are wet or dry. In any condition, the soil must be dense and strong enough to support the weight of the building. When soil is soft, the footing is made wider to spread the weight over more surface. In this way, each surface unit holds up less weight. Figure 3-6 shows how much weight various soil types will support. Standard footings should not be poured on loose soil.

Type of Soil	Bearing Capacity (pounds per ft²)
Soft Clay loose dirt, etc.	2 000
Loose Sand hard clay, etc.	4 000
Hard Sand or Gravel	6 000
Partially Cemented Sand or Gravel soft stone, etc.	20 000

Fig. 3-6 *Bearing capacity of typical soils.*

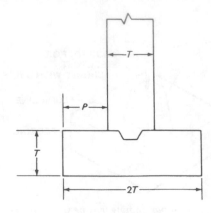

FLOORS	BASE-MENT	ALL WOOD FRAME		WOOD FRAME WITH MASONRY VENEER	
		T	P	T	P
1	None	6″	3″	6″	4″
	Yes	6″	2″	6″	3″
2	None	6″	3″	6″	4″
	Yes	6″	4″	8″	5″

NOTE: For soil with 2000 pounds per square foot (PSF) load capacity.

Fig. 3-7 *Typical footing size.*

Footing Width

The second factor is the width of the footing. As mentioned, the footing should be wider for soft soil. Figure 3-7 shows typical sizes for footings. As a rule, footings are about 2 times as wide as they are thick. The average footing is about 8 inches thick, and the footing is about the same thickness as the foundation wall.

Locating Footing Depth

Footings are laid out several inches below the frostline. For buildings with basements, place the top of the footing 12 inches below the frostline. For buildings that do not have basements, 4 to 6 inches below the frostline could be deep enough. Local building codes may give exact details.

Footings under Columns

The footings and foundations that most people see support only the outside walls. But today most houses are wide, and support is needed in the center of a wide building. This support is from footings, pillars, or columns built in the center. Pillars or columns must have a footing just as the outside walls do. For houses with basements, the footings and pillars become part of the basement floor and walls. See Fig. 3-8. Many houses do not have basements. Instead, they have a crawl space between the ground and the floor. This crawl space provides access to the pipes and utilities. Pillars or columns built on footings are used for supports in the crawl spaces. The footings may be any

Fig. 3-8 *Footings in a basement later became a part of the basement floor.*

Fig. 3-9 *Footings and piers must be located in crawl spaces.*

shape—square, rectangular, or round. Figure 3-9 shows a site prepared in this manner.

Footings for either basements or crawl spaces are all similar. They should be below the frostline as in a regular footing. However, they carry a greater weight than do the outside footings. For this reason, they should be 2 to 3 feet square.

Special Strength Needs

Footings for heavier areas of a building such as chimneys, fireplaces, bases for special machinery, and other similar things should be wider and thicker. For chimneys in a one-story building, the footing should project at least 4 inches on each side. The chimneys on two-story buildings are taller and heavier. Therefore, the footing should project 6 to 8 inches on each side of the chimney. Figure 3-10 shows a foundation for a fireplace.

Fig. 3-10 *A special footing is used for fireplaces. It supports the extra weight.*

Reinforcement and Strength

Two things are done to the footing to make it stronger. First, it is reinforced with steel rods. Then, the footing is also matched or keyed so that the foundation wall will not shift or slide.

Reinforcement In most cases the footing should be reinforced with rods. These reinforcement rods are called *rebar*. Two or more pieces of rebar are used. The rebar should be located so that at least 3 inches of space for concrete is left around all edges. See Fig. 3-11.

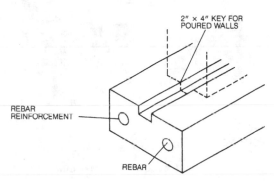

Fig. 3-11 *Footings may be reinforced. Note the key to keep the foundation from shifting.*

Keyed footings The best type of separate footing is keyed, as shown in Fig. 3-11. This means that the footing has a key or slot formed in the top. The slot is filled when the foundation is formed. The key keeps the foundation from sliding or moving off the footing. Without a key, freezing and thawing of water in the ground could force foundation walls off the footing.

EXCAVATING THE FOOTINGS

The procedure for locating the building on the lot was explained in Chapter 2. Batter boards were put up and lines were strung from them to show the location of the walls and corners.

Now the size, shape, and depth of footings must be decided.

Finding Trench Depth

Trenches or ditches must be dug, or excavated, for the footings and foundation. Ground that is extremely rough and uneven should be rough-graded before the excavation is begun. The topsoil that is removed can be piled at one edge of the building site. It should be used later when the ground is smoothed and graded around the building. Before the digging is started, determine how deep it is to be.

The trench at the lowest part of the site must be deep enough for the footing to be below the frostline. If the footing is to be 12 inches below the frostline, the trench at the lowest part must be deep enough for this. Figure 3-12 shows these depths. This lowest point becomes the level line for the entire footing. Elevations are taken at each corner to find out how deep the trenches are at each corner.

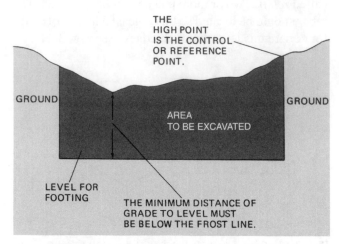

Fig. 3-12 *Footings must be below the frostline.*

Excavating for Deep Footings

Footings must be deep in areas where the frostline is deep. Deep footings are also needed when a basement will be dug.

Rough lines are drawn on the ground. They do not need to be very accurate, but the lines from the batter boards are used as guides. However, the rough line should be about 2 feet outside the line. See Fig. 3-13. The trench for the footing is dug much wider than the

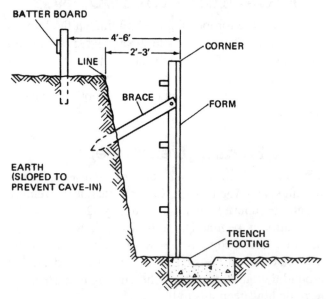

Fig. 3-13 *The trench is wider than the footing and sloped for safety.*

footing so that there is room to work. Since the footings are made of concrete, the molds (called *forms*) for the concrete must be built. Room is needed to build or put up the forms. Work that must be done after the footing or foundation is formed includes removing the forms, waterproofing the walls, and making proper drainage.

As the trench is dug, the depth is measured. When the trench has been dug to the correct depth, the machinery is removed. The forms are then laid out. For basements, the interior ground is also excavated.

Excavating for Shallow Footings

Rough lines are marked on the ground with chalk or shovel lines. These lines should be marked to show the width of the footing desired. Corner stakes are removed, and lines are taken from the batter boards. A trench is excavated to the correct depth.

The special footings for the interior are also excavated at this point. The excavation for the interior pad footings should be made to the same depth as that for the outside walls.

The concrete is poured directly into these trenches. Any reinforcement is made without forms in the excavation itself. The rebar is suspended with rebar stakes or metal supports called *chairs*. See Fig. 3-14A through C.

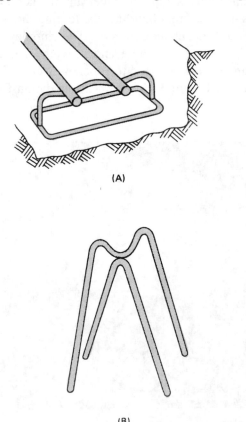

Fig. 3-14 *Chairs are used to hold rebar in place.* (Richmond Screw Anchor)

SYMBOL	BAR SUPPORT ILLUSTRATION	BAR SUPPORT ILLUSTRATION PLASTIC CAPPED OR DIPPED	TYPE OF SUPPORT	SIZES
SB		CAPPED	Slab Bolster	¾, 1, 1½, and 2 inch heights in 5 ft. and 10 ft. lengths
SBU*			Slab Bolster Upper	Same as SB
BB		CAPPED	Beam Bolster	1, 1½, 2, over 2" to 5" heights in increments of ¼" in lengths of 5 ft.
BBU*			Beam Bolster Upper	Same as BB
BC		DIPPED	Individual Bar Chair	¾, 1, 1½, and 1¾" heights
JC		DIPPED DIPPED	Joist Chair	4, 5, and 6 inch widths and ¾, 1 and 1½ inch heights
HC		CAPPED	Individual High Chair	2 to 15 inch heights in increments of ¼ inch
HCM*			High Chair for Metal Deck	2 to 15 inch heights in increments of ¼ in.
CHC		CAPPED	Continuous High Chair	Same as HC in 5 foot and 10 foot lengths

(C)

Fig. 3-14 *Types and sizes for bar supports. (Continued)*

To form a key in this type of footing, stakes are driven along the edges as in Fig. 3-15. The board that forms the key in the footing is suspended in the center of the trench area.

In many areas, concrete block wall is used on this type of footing. The blocks may be secured by inserting rebars into the footing area. The bricks or blocks are laid so that the rebar is centered in an opening in the block. The opening is then filled with mortar or concrete to secure the foundation against slipping.

The important thing to remember is that special forms are not used with shallow footings. Also, they may not be finished smooth. As a result they may appear very rough or unfinished. This is not important if they are the proper shape.

Slab Footings and Basements

Slab footings are used in areas where concrete floors are made. Slabs can combine the concrete floor and the footing as one unit. Slabs, basement floors, and other large concrete surfaces are detailed in Chapter. 4. Basement floors are made separately from the footings and are done after the footings and basement walls are up.

BUILDING THE FORMS FOR THE FOOTINGS

After the excavation has been completed, the corners must be relocated. After the corners are relocated, the forms are built and leveled and the concrete is poured and allowed to harden. Then the forms are removed, and the foundation is erected. In many cases, the footing and the foundation are made as one piece.

Laying Out the Forms

After the excavation is complete, the first step is to relocate the corners and edges for the walls. To do this, the lines from the batter boards are restrung and a plumb bob is used to locate the corner points. The corner points and other reference points are marked with a stake. The stake is driven level with the top of the footing. This level

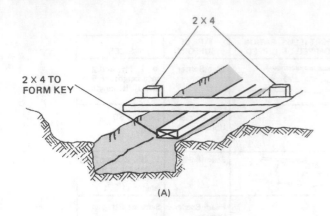

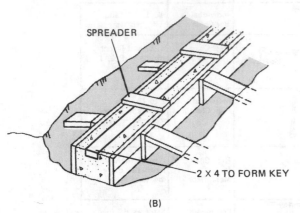

Fig. 3-15 *Keys are made by suspending 2 × 4s in the form. (A) Keys for trench forms. (B) Keys for board forms.*

is established by using a transit or a level. Refer to Chap. 2 for this procedure.

Nails

It is best to use double-headed, or duplex, nails for making the forms. Forms should be nailed with the nails on the outside. This means that the nails are not in the space where the concrete will be. This way the nail head does not get embedded in the concrete and is left exposed. The double head allows the nail to be driven up tight; it will still be easy to pull out when the forms are taken apart.

Putting Up the Forms

With the corner stakes used for location and level, the walls of the forms are constructed. The amount that the footing is to project past the wall is determined. Usually this is one-half the thickness of the foundation wall. This dimension is needed because the corner indicates foundation corner and not footing corner. Stakes are driven outside the lines so that the form will be the proper width. The carpenter must allow for the width of the stake and the width of the boards used for the forms. See Fig. 3-16. Drive stakes as needed for support. As a rule, the distance between stakes is about twice the width of the footing. Nail the top board to the first stake and level the top board in two directions. For the first direction, the top board is leveled with the corner stake. For the second direction, the top board is leveled on its length. See Fig. 3-17. After the top board is leveled, nail it to all stakes. Then nail the lower boards to the stakes. Both inside and outside forms are made this way.

If 1-inch-thick boards are used to build the form, stakes should be driven closer together. If boards 2 inches thick are used, the stakes may be 4 to 6 feet apart. In both cases, the stakes are braced as shown in Fig. 3-18.

Loose dirt should be removed from under the footing form. It is best for the footing to be deeper than is needed. Never make a footing thinner than the specifications. Never fill any irregular hole or area with loose

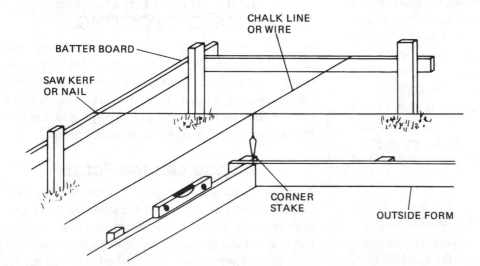

Fig. 3-16 *The footing corner is located and leveled.*

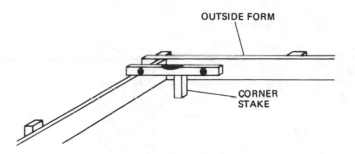

Fig. 3-17 The form is leveled all around.

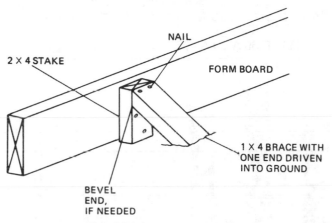

Fig. 3-18 Bracing form boards.

dirt. Always fill with gravel or coarse sand and tamp it firmly in place.

The keyed notch The key or slot in the footing is made with a board. The board is nailed to a brace that reaches across the top of the forms. The brace should be nailed in place at intervals of 4 feet apart. Refer to Fig. 3-15 to see how the key is made.

Excavation for drains and utility lines Drainpipes and utility lines are sometimes located beneath the footings in a building. When this is done, trenches are dug underneath the footing forms. These trenches are usually dug by hand underneath the forms. After the drainpipes or utility lines are laid in place, the area is filled with coarse gravel or sand. This gravel or sand is tightly packed in place beneath the form.

Spacing the walls of the form The weight of the concrete can make the walls spread apart. To keep the walls straight, braces are used. The braces on the walls provide much support. Special braces called *spreaders* are also nailed across the top. Forms should be braced properly so that the amount of concrete ordered will fill the forms properly. Also, this practice ensures that excess concrete does not add extra weight to the building.

The forms should also be checked to make sure that there are no holes, gaps, or weak areas. These could let the concrete leak out of the form and thus weaken the structure. These leaks are called *blowouts*.

WORKING WITH CONCRETE

Before the concrete is ordered and poured, several things are done. The forms should be checked for the proper depth and level. Openings and trenches beneath the footing area for pipes and utility lines are made. These should be properly leveled and filled. The forms should be checked to make sure that they are properly braced and spaced. Finally, chalk lines and corner stakes should be removed from the forms.

Reinforcement

In most cases, reinforcement rods (rebar) are placed in the footing after the forms are finished. The amount of reinforcement is usually given in the plans. As it is laid, the rebar is tied in place. Soft metal wires, called *ties*, are twisted around the rebars. The carpenter must be sure that the footing conforms to the local building codes. See Fig. 3-19 and Table 3-1.

Specifying Concrete

Most concrete used today is made from cement, sand, and gravel mixed with water. The cement is the "glue" that hardens and holds or binds the materials. Most cement used today is portland cement. It is made from limestone that is heated, powdered, and mixed with certain minerals. When mixed with aggregates, or sand and gravel, it becomes concrete.

Concrete mixes can be denoted by three numbers, such as 1–2–3. This is the volume proportion of cement, sand, and gravel. The 1–2–3 is the basic mix, but it is varied for strength, hardening speed, or other factors. However, it is recommended that concrete be specified by the water-to-cement ratio, aggregate size, and bags of concrete per cubic yard. See Table 3-2.

Most concrete today is delivered to the building site. Usually, the concrete is not mixed by the carpenters. It is delivered by concrete trucks from a concrete company. Figure 3-20 shows a transit-mix truck. The concrete is sold in units of cubic yards. The carpenter may need to make the order for concrete to the concrete company. To do so, the carpenter must be able to figure how much concrete to order.

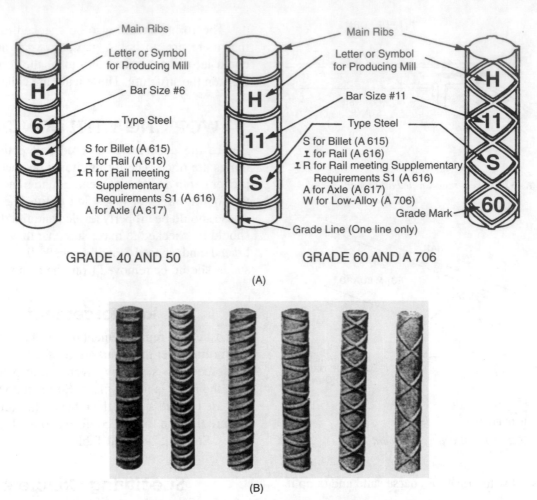

Fig. 3-19 *(A) Identification marks for ASTM standard reinforcing bars. (B) Various types of deformed bars.*

Table 3-1 *ASTM Standard Reinforcing Bars*

Bar Designation No.	Nominal Weight, lb/ft	Nominal Dimensions		
		Diameter, in	Cross-Sectional Area, in	Perimeter, in
3	0.376	0.375	0.11	1.178
4	0.668	0.500	0.20	1.571
5	1.043	0.625	0.31	1.963
6	1.502	0.750	0.44	2.356
7	2.044	0.875	0.60	2.749
8	2.670	1.000	0.79	3.142
9	3.400	1.128	1.00	3.544
10	4.303	1.270	1.27	3.990
11	5.313	1.410	1.56	4.430
14	7.65	1.693	2.25	5.32
18	13.60	2.257	4.00	7.09

Setting Time

Setting and hardening of a concrete are a continuous process. However, two points are important to consider.

- Initial setting time is the interval between the mixing of the concrete with water and the time when the mix has lost plasticity and has stiffened to a certain

degree. It marks roughly the end of the period when the wet mix can be molded into shape.

- Final setting time is the point at which the set concrete has acquired a sufficient firmness to resist a certain pressure. Most specifications require an initial minimum setting time at ordinary temperatures

Table 3-2 Concrete Use Chart

Uses	Concrete, Bags per Cubic Yard	Sand, Pounds per Bag of Concrete	Gravel, Pounds per Bag of Concrete	Gravel Size, Average Diameter in Inches	Water, Gallons per Bag of Concrete	Consistency Slump
Footings, basement walls (8-inch), or foundation walls (8-inch thickness)	5.0	265	395	1¹/₂″	7	4 – 6 inches
Slabs, basement floors, sidewalks, etc. (4-inch thickness)	6.2	215	295	1″	6	4 – 6 inches
Basic 1·2·3 mixture (approximation only)	6.0	190	275	2″	5.5	2 – 4 inches

NOTES: 1. All figures are for slight to moderate ground water and medium-fineness sand.
2. All figures vary slightly.

Fig. 3-20 *A transit-mix truck delivers concrete to a site.*

of about 45 minutes and a final setting time of no more than 10 to 12 hours.

To reach its full strength, it takes concrete 28 days.

Estimating Concrete Needs

A formula is used to estimate the volume of concrete needed. The basic unit for concrete is the cubic yard. A cubic yard is made up of 27 cubic feet ($3 \times 3 \times 3$). To convert footing sizes, use the formula

$$\frac{L'}{3} \times \frac{W''}{36} \times \frac{T''}{36} = \text{cubic yards}$$

where L' = length in feet
W'' = width in inches
T'' = thickness in inches

Example. A footing is 18 inches wide and 8 inches thick. It must support a building 48 feet long and 24

feet wide. The distance around the edges is called the perimeter. The perimeter is $(2 \times 48) + (2 \times 24)$, or 144 feet. This would be

$$\frac{L'}{3} \times \frac{W''}{36} \times \frac{T''}{36} = \text{cubic yards}$$

$$= \frac{144}{3} \times \frac{18}{36} \times \frac{8}{36}$$

and by cancellation,

$$\frac{48}{1} \times \frac{1}{2} \times \frac{2}{9} + \frac{96}{18} = 5.33 \text{ cubic yards}$$

The minimum amount that can be ordered is 1 cubic yard. After the first cubic yard, fractions can be ordered. The estimate is 5⅓ cubic yards. Often a little more is ordered to make sure enough is delivered.

Pouring the Concrete

To be ready, the carpenter sees to two things. First the forms must be done. Then the concrete truck must have a close access. The driver can move the spout to cover some distance. However, it may be necessary to carry the concrete an added distance. This can be done by pumping the concrete or by carrying it. Wheelbarrows, as in Fig. 3-21, are sometimes used.

Another method is to use a dump bucket carried by a crane. See Fig. 3-22.

The builder must spread, carry, and level the concrete. The truck will only deliver it to the site. The truck driver can remain only a few minutes. The driver is not allowed to help work the concrete. As the concrete is poured, it should be tamped. This is done with a board or shovel that is plunged into the concrete. See Fig. 3-23. Tamping helps get rid of air pockets. This makes the concrete solidly fill all the form.

Fig. 3-21 *Sometimes the concrete must be carried from the truck to the work site.* (IRL Daffin)

Fig. 3-22 *A dump bucket is used to dump concrete into forms that trucks can't reach* (Universal Form Clamp)

For shallow footings, no smoothing or "finishing" need be done. For deep footings, the surface should be roughly leveled. This is done by resting the ends of a board across the top of the form. The board is then used to scrape the top of the concrete smooth and even with the form.

Strength of the Concrete

The tests that measure the rate at which concrete develops strength are usually made on a mortar commonly composed of 1 part cement to 3 parts sand, by weight, mixed with a defined quantity of water. Tensile tests on briquettes, shaped like a figure 8, thicker at the center,

Fig. 3-23 *Concrete is tamped into forms to get rid of air pockets.* (Portland Cement)

were formerly used, but have been replaced or supplemented by compressive tests on cubical specimens or traverse tests on prisms. The American Society for Testing and Materials (ASTM) specification requires tensile tests on a 1:3 cement-sand mortar and compressive tests on a 1:2.75 mortar. In all these tests, the size grading of the sand and usually its source are specified.

Most testing of concrete requires minimum strength at 3 and 7 days, and sometimes 28 days is specified. However, for rapid-hardening portland cement, a test at 1 day is sometimes required. Cement with high alumina content requires tests at 1 and 3 days. For most residential uses, the *slump test* is sufficient for those working in the field. It is visual and rather easy to perform.

Figure 3-24 shows how the slump test is performed, and Table 3-3 shows the allowable slump for various uses.

Slump Test

The slump test is an easy test to perform. Just follow the six steps outlined below.

1. Fill one-third of the cone with concrete and rod 25 times (Fig. 3-24A).

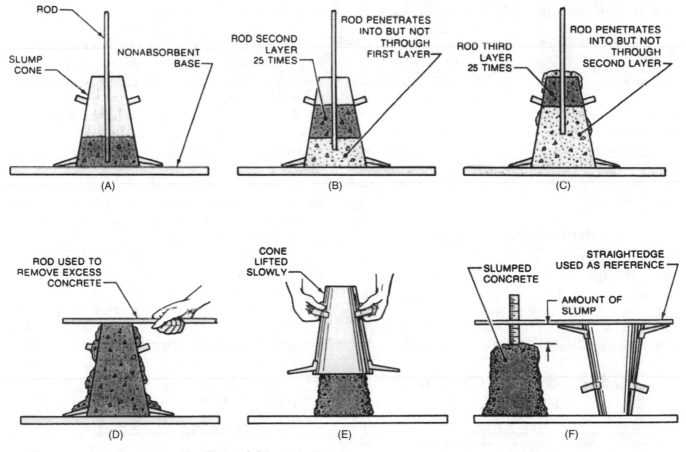

Figure 3-24 *Slump test. (Courtesy of Portland Cement Association)*

Table 3-3 *Allowable Slump*

Concrete Construction	Slump, Inches	
	Max.*	Min.
Reinforced foundation walls and footings	3	1
Plain footings, caissons, and substructure walls	3	1
Beams and reinforced walls	4	1
Building columns	4	1
Pavements and slabs	3	1
Mass concrete	2	1

* May be increased 1 inch for consolidation by hand methods such as rodding and spading. *(Portland Cement Association)*

2. Fill two-thirds of the cone and rod the second layer 25 times (Fig. 3-24B).

3. Fill the cone to overflow and rod 25 times (Fig. 3-24C).

4. Remove the excess from the top of the cone and at the base (Fig. 3-24D).

5. Lift the cone vertically with a slow, even motion (Fig. 3-24E).

6. Invert the cone and place it next to the concrete. Measure the distance from the top of the cone to the top of the concrete (Fig. 3-24F).

Effects of Temperature

The hardening of concrete speeds up if its temperature is warm and slows down when cool. If a concrete mix begins to set in 2 hours at 70°F, it may set in 1 hour or less at 95°F. For the same mix, the setting time may increase to 3 hours or more at 50°F and 5 hours or longer at 35°F.

Temperature also affects the amount of water needed to make a cubic yard of workable concrete. At low temperatures, less water is needed to get a workable mix than at high temperatures. For typical concrete mixes having a given slump, an additional gallon of water per cubic yard of concrete is needed for each 12°F increase in its temperature. This means that the same amount of cement is used per cubic yard of concrete. The water-cement ratio will be higher for warm concrete than for cold concrete with the same slump.

The best temperature for making concrete depends on how long the concrete will be cured. If concrete will be cured naturally for long periods such as in dams or foundations, concrete made at cold temperatures or those just above freezing will eventually be stronger than concrete made at higher temperatures. Concrete that can be cured only a few days, however, should be kept above 60°F. The heat of hydration generated by the setting action of cement also affects the concrete strength gain.

Concrete that freezes soon after it is made and before its strength reaches about 500 pounds per square inch may be permanently damaged. Frozen concrete must be removed and replaced. Admixes do not lower the freezing point of concrete significantly so admixtures do not behave as antifreeze agents. Accelerating admixtures will speed up hardening and reduce the time required to reach a strength of 500 pounds per square inch.

Curing of the Concrete

The quality of the curing affects the strength, durability, and other qualities. Curing is done by keeping moisture in concrete long enough for most of the cement to hydrate. The time depends on the temperature of the concrete. An early start is just as important as the temperature and duration of the curing.

Using Portland Cement Concrete Safely

Fresh concrete made with portland cement is highly alkaline or caustic and can cause skin irritation and burns. Follow these simple precautions to avoid needless injury.

- *Keep cement products off the skin.* Experienced concrete craftsmen protect their skin with boots, gloves, clothing, and kneepads. Skin injury may result from clothing that is wet from cement mixtures.

- *Don't let the skin rub against cement products.* Many cement products are abrasive. Rubbing increases the chances of serious injury.

- *Wash the skin promptly after contact with cement products.*

- *Keep cement and cement products out of the eyes.* Many construction workers use safety glasses. If any cement or cement mixtures get into the eye, flush immediately and repeatedly with water. Consult a physician promptly.

- *Keep products out of the reach of children.* Keep children away from cement powder and all freshly mixed cement products.

BUILDING THE FOUNDATION FORMS

The foundation is a wall between the footing and the floor of the building. It is often made of concrete. However, it may also be made of concrete blocks, bricks, or stone. In some regions, foundation walls are made of treated plywood as well.

When concrete is used, special forms may be used for the foundations. These forms are easily put up and taken down. Often the footing and the foundation are made in one solid piece.

Builders also make the foundation in much the same way as they make the forms for the footings. After the form is removed, the lumber is used in framing the house.

In either method, the form should be spaced for the correct width. It must also be spaced to prevent the weight of the concrete from spreading the forms. The width of the form is important. A form that is not wide enough will not carry the weight of the building safely. A form that is too wide uses too much concrete. Too much concrete costs more and adds weight to the building. This weight can cause settling problems. However, spreading forms can also cause errors in pouring the concrete. Frequently, just enough concrete is ordered to fill the forms. If the forms are allowed to spread, more concrete is used. Thus enough concrete might not be delivered to the site.

Making the Forms

In making forms, several things must be considered. First, sections of a form must fit tightly together. This prevents leaks at the edges. Leaks can cause bubbles and air pockets in the concrete. This is called *honeycomb*. Honeycomb weakens the foundation wall.

When special forms are used and assembled to make the total form, they must be braced properly. Forms up to 4 feet wide are braced on the back side with studs. These forms are made from metal sheets or from plywood sheathing ¾ inch thick or thicker. For building walls higher than 4 feet, special braces called *wales* are used. See Fig. 3-25.

The sheathing is nailed to the studs and wales from the inside. The studs are laid out flat on the ground. The sheathing is then laid on the studs and nailed down. The assembled form is then erected and placed into position. It is spaced properly, and wales and braces are added.

Braces are erected every 4 to 6 feet. However, for extra weight or wall height, braces may be closer.

Joining the Forms

Edges and corners should be joined tightly so that no concrete leaks occur. When you are using plywood forms, join the edges by nailing the plywood sheathing to the studs. Use 16d nails as in Fig. 3-26. When you nail

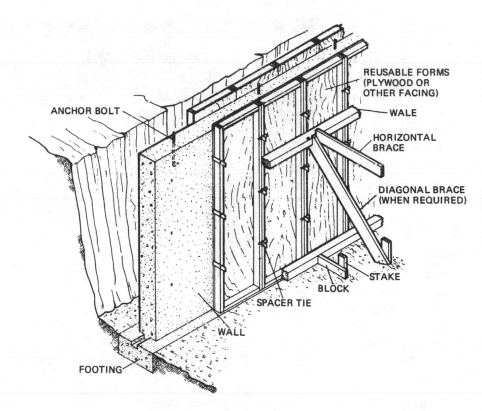

Fig. 3-25 *Special braces called wales are sometimes needed.* (Forest Products Laboratory)

Labels: ANCHOR BOLT, REUSABLE FORMS (PLYWOOD OR OTHER FACING), WALE, HORIZONTAL BRACE, DIAGONAL BRACE (WHEN REQUIRED), STAKE, BLOCK, SPACER TIE, WALL, FOOTING

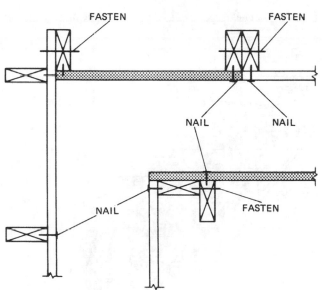

Fig. 3-26 *Nailing plywood panel forms.*

Labels: FASTEN, FASTEN, NAIL, NAIL, NAIL, FASTEN

the corners together, use the procedure also shown in Fig. 3-26. Again, 16d nails are used.

When special metal forms are used, the manufacturer's directions should be carefully followed.

Spreaders

Spreaders are used on all forms to hold the walls apart evenly. Several types may be used. The spreaders may be made of metal at the site. Metal straps may be nailed in between the sections of the forms. However, most builders use spacers that have been made at a factory for that type of form.

After the concrete has hardened, the forms are removed. The spreader is broken off when the form is removed. Special notches are made in the rods to weaken them so that they will break at the required place. Figure 3-27 shows a typical spreader.

Using Panel Forms

A panel form is a special form made up in sections. The forms may be used many times. Most are specially made by manufacturers. Each style has special connectors that enable the forms to be quickly and easily erected. By use of standard sizes such as 2- × 4- or 4- × 8-foot sections, walls of almost any size and shape

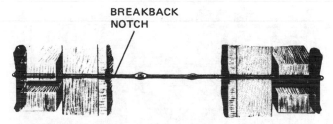

Fig. 3-27 *Special braces called spreaders keep the forms spaced apart. These rods are later broken off at the notches, called breakbacks. The rest of the spreader remains in the concrete.* (Richmond Screw Anchor)

Label: BREAKBACK NOTCH

can be erected quickly. The forms are made of metal or wood. The advantages are that they are quick and easy to use, they may be used many times, and they may be used on almost any size or shape of wall.

Panel forms must be braced and spaced just as forms constructed on the site are. To the builder making one building, there is little advantage to using such forms. They must be purchased, and they are not cheap. However, when they are used many times, the savings in time make them economical. Figure 3-28 shows an example.

One-Piece Forms

When the same style of footing and foundation is used in the same type of soils, a one-piece form is used. This combines the footings and the foundation as one piece (Fig. 3-29). Several versions may be used. Some types allow a footing of any size to be cast with a foundation wall of any thickness. Some incorporate the footing and the foundation wall as a stepped figure. Others, as in Fig. 3-30, use a tapered design.

The one-piece form saves operational steps. Casting is quicker, it is easier, and it is done in one operation. Two-piece forms and the conventional processes require that the footing be cast and allowed to harden. The footing forms are then removed and the founda-

Fig. 3-29 *Panel forms can combine the footing and the foundation.* (Proctor Products)

Fig. 3-30 *Tapered form.* (Proctor Products)

Fig. 3-28 *Reusable forms are made from panels of plywood or metal. These special panel forms are assembled with special fasteners.* (Proctor Products)

tion forms erected, cast, allowed to harden, and removed. This takes several days and many hours of work. The one-piece form offers many advantages in the savings of time and cost of labor.

Special Forms

Certain types of form are used for shapes that are commonly used. A round form such as that in Fig. 3-31 is commonly used. This type of form may be used to cap pilings. It may also be used for the footings under the central foundation pillars.

Fig. 3-31 *Pier form. (Proctor Products)*

Other special forms include forms made of steel and cardboard. Steel forms may be used to cast square or round columns. They are normally used in construction of large projects such as bridges and dams. They are also used on large business buildings. The cardboard forms are made of treated paper and fibers. They are used one time and destroyed when they are removed. Figure 3-32 shows a pillar made with a cardboard form.

Fig. 3-32 *A round form made of cardboard. (Proctor Products)*

All such special forms allow time and labor to be saved. Little labor and time is needed to set up special forms. Reinforcement may be added as required. Also, such forms are available in many shapes and sizes.

Openings and Special Shapes

Openings for windows and doors are frequently required in concrete foundation walls. Also, special keys or notches are often needed. These hold the ends of support frames, joists, and girders. At times, utility and sewer lines run through a foundation. Special openings must also be made for these. It is very expensive to try to cut such openings into concrete once it is hardened and cured.

However, if portions of the forms are blocked off, concrete cannot enter these areas. This way almost any shape can be built into the wall before it is formed. This shape is called a *block-out* or a *buck*. The concrete is then poured and moves around these blocked-out areas. When the concrete hardens, the shape is part of the wall. This is quicker, cheaper, and easier. It also provides better strength to the wall and makes the forms used more versatile.

Of course, a carpenter can build a block-out of almost any shape in a form. First, one wall of the form is sheathed. The shape can then be framed out on that side. The inside of the shaped opening may be used for nails and braces. The outside, next to the concrete, should be kept smooth and well finished. However, it is expensive to pay a carpenter to frame special openings if they have to be repeated many times. It is better to use a form that can be used over again. Figure 3-33 shows an example.

Fig. 3-33 *Openings may be made with special forms that can be reused. (Proctor Products)*

When you are building a buck or block-out, first check the plans. Sometimes bucks are removed; other times they are left to form a wooden frame around the opening.

In either case, the size of the opening is the important thing. To determine rough opening sizes for windows in walls, see Building Walls in *Miller's Guide to Framing and Roofing*.

Buck keys Strips of wood are used along the sides of openings in concrete. These are used as a nail base to hold frames or units in the opening. These strips are called *keys*. See Fig. 3-34. If the buck is removed, the key is left in place. If the buck is left in place, the key holds the buck frame securely. Note how the key is undercut. The undercut prevents the key from being pulled out.

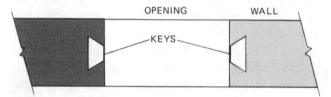

Fig. 3-34 *Keys are placed along the sides of openings as a nail base. Note undercut so that key cannot be removed.*

Bucks should be made from 2-inch lumber. The key can be made of either 1- or 2-inch lumber. The key needs to be only 1 or 2 inches wide. Usually, only the sides of the bucks are keyed.

Buck left as a frame First find the size of the opening. Next, cut the top and bottom pieces longer than the width. These pieces are usually 3 inches longer than the width. See Fig. 3-35. Next, cut the two sides to the same height as the desired opening. Nail with two or three 16d nails as shown. Note that the top piece goes over the sides. The desired size for the opening is the same as the size of the opening in the buck.

Buck removed First find the opening size. Next, cut the top and bottom pieces to the exact width of the opening. Then cut the two sides shorter. The amount is usually 3 inches (twice the thickness of the lumber used). See Fig. 3-36. Nail the frame together with 16d nails as shown.

Note that the opening size is the same as the outside dimensions of the buck. Also, the outside faces are oiled. This keeps the concrete from sticking to the sides of the buck. It also makes it easier to remove the buck.

Buck braces When the opening is large, the weight of the concrete can bend the boards. If the boards bend, the opening will not be the right size or shape.

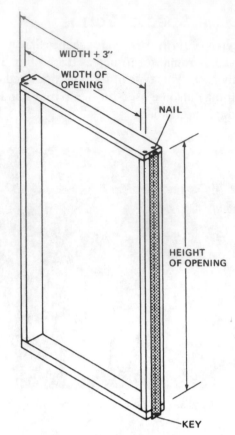

Fig. 3-35 *The buck may be left in the wall as the frame. In this case the opening in the buck is the desired size.*

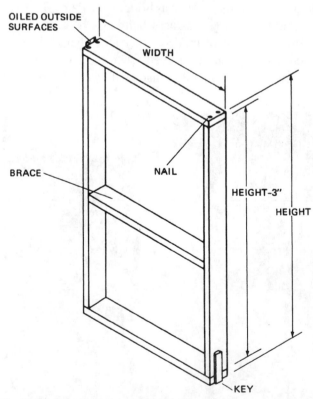

Fig. 3-36 *The buck may be removed. In this case the buck frame is the size desired.*

To prevent this, braces are placed in the opening. See Fig. 3-36. Note that the braces can run from side to side or from top to bottom.

Another type of form is used for porches, sidewalks, or overhangs. See Fig. 3-37A to D. This allows porch supports to be part of the foundation. The earth is filled in later.

Fig. 3-37A *Special forms used for an overhang on a basement wall.*

Reinforcing Concrete Foundations

Concrete is very strong when compressed. However, it does not support weight without cracking. Even though it is very hard, it is also very brittle. In order to resist shifting soil, concrete should be reinforced. It is not a matter of whether the soil will shift. It is more a matter of how much the soil will shift.

It should be noted that sometimes the reinforcement is added before the forms are done. When the forms are tall, very narrow, or hard to get at, reinforcement is done first.

Concrete reinforcement is done in two basic ways. The first way is to use concrete reinforcement bars. The second way is to use mesh. Mesh is similar to a large screen made with heavy steel wire. The foundations

Fig. 3-37B *The overhang after the forms are removed.*

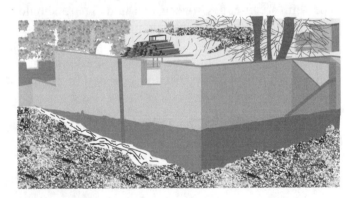

Fig. 3-37C *Basement walls are coated and waterproofed.*

Fig. 3-37D *Finally, the concrete porch will be poured.*

may be reinforced by running rebars lengthwise across the top and at intervals up and down. Figure 3-38 shows some typical reinforcement.

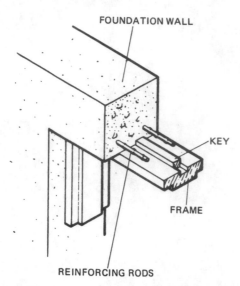

FOUNDATION WALL

KEY

FRAME

REINFORCING RODS

Fig. 3-38 *Foundations should be reinforced at the top.* (Forest Products Laboratory)

The amount and type of reinforcement used are determined by the soil and geographic location. The reinforcement is spaced and tied. There should be at least 3 inches of concrete around the reinforcement.

The carpenter should be sure that the foundation and reinforcement conform to the local building codes.

Estimating Concrete Volume

When the forms are complete, the amount of concrete needed is computed. The same formula as for footings is used:

$$\frac{L'}{3} \times \frac{W''}{36} \times \frac{T''}{36} = \text{cubic yards}$$

However, W in inches is replaced by the wall height H in feet, so that

$$\frac{L'}{3} \times \frac{H'}{3} \times \frac{T''}{36} = \text{cubic yards}$$

If the foundation is to be 8 inches thick and 8 feet high and the perimeter is 144 feet, then

$$\frac{144}{3} \times \frac{8}{3} \times \frac{8}{36} = \text{cubic yards}$$

and by cancellation,

$$\frac{48}{1} \times \frac{8}{3} \times \frac{2}{9} = \frac{16}{1} \times \frac{8}{1} \times \frac{2}{9}$$

$$= \frac{256}{9}$$

$$= 28.4 \text{ cubic yards}$$

Delivery and Pouring

Once the needs are estimated and the concrete has been ordered or mixed, the wall should be poured. If a transit-mix truck is used, the concrete is mixed and delivered to the site. The concrete truck should be backed as close as possible to the forms. As in Fig. 3-39, the concrete should be poured into the forms, tamped, and spread evenly. By doing this, air pockets and honeycombs are avoided.

Fig. 3-39 *Concrete is poured or pumped into the finished form.*

Finishing the Concrete

Two steps are involved in finishing the pouring of the concrete foundation wall. First, the tops of the forms must be leveled. Sometimes the concrete is poured to within 2 or 3 inches of the top. The concrete is then allowed to partially cure and harden. A concrete or grout with a finer mixture of sand may be used to finish out the top of the foundation.

Anchors are embedded in the concrete before it hardens completely. One end of each anchor bolt is threaded and the other is bent. See Fig. 3-40. The threaded end sticks up so that a sill plate may be bolted in place. As the concrete begins to harden, the bolts are slowly worked into place by being twisted back and forth and pushed down. Once they are embedded firmly, the concrete around them is troweled smooth. Figure 3-41 shows an anchor embedded in a foundation wall.

The forms are removed after the concrete cures and hardens. Low spots must be filled. A small spot can be shimmed with a wooden shingle. However, a larger area should be filled in with grout or mortar.

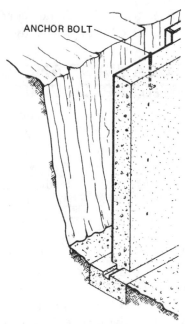

Fig. 3-40 *Anchor bolt.*

Fig. 3-41 *Anchor bolts are embedded in foundation walls and protrude from them.* (Forest Products Laboratory)

CONCRETE BLOCK WALLS

Concrete blocks are often used for basement and foundation walls. When the foundation is exposed, it may be faced with brick. See Fig. 3-42. Blocks need no form work and go up more quickly than brick or stone. The most common size is 7⅝ inches high, 8 inches wide, and 15⅝ inches long. The mortar joints are ⅜ inch wide. This gives a finished block size of 8 × 16 inches, and a wall 8 inches thick.

Fig. 3-42 *Concrete block foundations may be faced with brick or stone.*

The footing is rough and unfinished. This is so because the mortar for the block is also used to smooth out the rough spots. No key is needed, but reinforcement rod should be used. Figure 3-43 shows a footing for a block wall.

Block walls should be capped with either concrete or solid block. Anchor bolts are mortared in the last row of hollow block. They then pass through the mortar joint of the solid cap. See Fig. 3-44.

A special pattern is sometimes used to lay the block. This is done when the wall will also be the visible finish wall. This pattern (see Fig. 3-44) is called a *stack bond*. It should be reinforced with small rebar.

PLYWOOD FOUNDATIONS

Plywood may be used for foundations or for basement walls in some regions. There are several advantages to using plywood. First, it can be erected in even the coldest weather. It is fast to put up because no forms or

Fig. 3-43 *A footing for a concrete block foundation.*

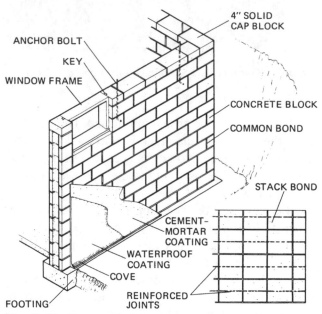

Fig. 3-44 *Concrete block basement wall.* (Forest Products Laboratory)

Labels for Fig. 3-44:
ANCHOR BOLT
KEY
WINDOW FRAME
4" SOLID CAP BLOCK
CONCRETE BLOCK
COMMON BOND
STACK BOND
CEMENT-MORTAR COATING
WATERPROOF COATING
COVE
FOOTING
REINFORCED JOINTS

reinforcement is required. For the owner, plywood makes a wall that is warmer and easier to finish inside. It also conserves on the energy required to heat the building.

The frame is formed with 2-inch studs located on 12- or 16-inch centers. The frame is then sheathed with plywood. Insulation is placed between the studs. The

exterior of the wall is covered with plastic film, and building paper is lapped over the top part. The building paper is laid over the top of the rock fill and helps drainage. See Figs. 3-45 and 3-46.

TYPICAL PANEL

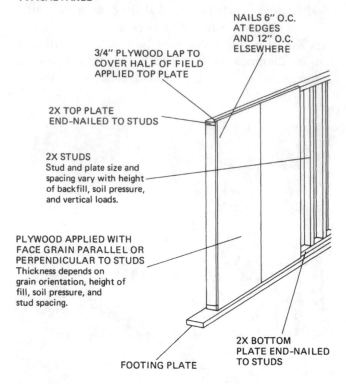

Labels for Fig. 3-45:
NAILS 6" O.C. AT EDGES AND 12" O.C. ELSEWHERE
3/4" PLYWOOD LAP TO COVER HALF OF FIELD APPLIED TOP PLATE
2X TOP PLATE END-NAILED TO STUDS
2X STUDS
Stud and plate size and spacing vary with height of backfill, soil pressure, and vertical loads.
PLYWOOD APPLIED WITH FACE GRAIN PARALLEL OR PERPENDICULAR TO STUDS
Thickness depends on grain orientation, height of fill, soil pressure, and stud spacing.
2X BOTTOM PLATE END-NAILED TO STUDS
FOOTING PLATE

NOTE: Wood and plywood are treated.

Fig. 3-45 *A typical plywood foundation panel.* (American Plywood Association)

It is important to note that all lumber and plywood must be pressure-treated with preservative.

DRAINAGE AND WATERPROOFING

A foundation wall should be drained and waterproofed properly. If a wall is not drained properly, the water may build up and overflow the top of the foundation. Unless the wall is waterproofed, water may seep through it and cause damage to the foundation and footings. It may also cause damage to the interior of a basement. Proper drainage is ensured by placing drain pipes or drain tile around the outside edges of the footing. A gravel fill is used to place the tile slightly below the level of the top of the footing. Figure 3-47 shows the proper location of the drainpipe. The pipe is then covered with loose gravel and compacted slightly.

If the house has a basement, the foundation walls should be waterproofed. If the house has only a crawl space, no waterproofing is needed.

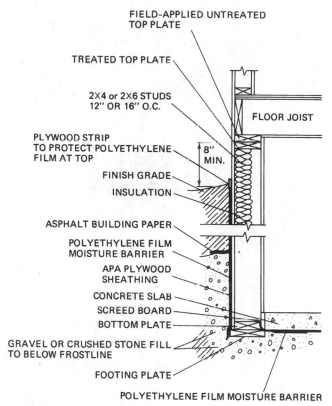

FIELD-APPLIED UNTREATED TOP PLATE

TREATED TOP PLATE

2X4 or 2X6 STUDS 12″ OR 16″ O.C.

FLOOR JOIST

PLYWOOD STRIP TO PROTECT POLYETHYLENE FILM AT TOP

8″ MIN.

FINISH GRADE

INSULATION

ASPHALT BUILDING PAPER

POLYETHYLENE FILM MOISTURE BARRIER

APA PLYWOOD SHEATHING

CONCRETE SLAB

SCREED BOARD

BOTTOM PLATE

GRAVEL OR CRUSHED STONE FILL TO BELOW FROSTLINE

FOOTING PLATE

POLYETHYLENE FILM MOISTURE BARRIER

Fig. 3-46 *Cross section of plywood basement wall.* (American Plywood Association)

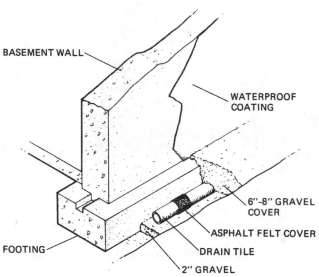

BASEMENT WALL

WATERPROOF COATING

6″-8″ GRAVEL COVER

ASPHALT FELT COVER

DRAIN TILE

2″ GRAVEL

FOOTING

Fig. 3-47 *Basement walls should be coated and drained.* (Forest Products Laboratory)

Waterproofing Basement Walls

Three types of walls are commonly used today for basements. The most common is the solid cast concrete wall. However, concrete blocks are also used and so is plywood.

Concrete block foundations Concrete blocks should be plastered and waterproofed. Figure 3-44 shows the processes involved. First, the concrete wall is coated with a thin coat of plaster. This is called a *scratch coat.* After this has hardened, the surface is scratched so that the next coat will adhere more firmly. The second coat of plaster is applied thickly and smoothly over both the wall and the top of the footing. As shown in Fig. 3-44, this outside layer is then covered with a waterproof coating. Such a coating could include layers of bitumen, builder's felt, or plastic.

Waterproofing concrete walls No plaster is needed over a cast concrete wall. The wall may be quickly coated with bitumen. However, plastic sheeting may also be applied to cover both the footing and the foundation in one piece. The most common process, as shown in Fig. 3-47, involves a bitumen layer. This bitumen layer is sometimes reinforced by a plastic panel which is then coated with another layer of bitumen.

Basement Wall Coatings

For many years, the asphalt-based coatings have been applied to basement walls to waterproof them. However, this coating is often damaged with the backfill and careless workmanship on the part of those responsible for cleaning up the work site. A more permanent and damage-resistant coating has now become commonly available. Rub-R-Wall foundations are coated with a 100% polymer membrane that is guaranteed not to leak for a lifetime. Conventional asphalt-based coatings, such as cutback or emulsified asphalts, are susceptible to leaching. The Rub-R-Wall polymer membrane is virtually leach proof. Because of its nonhazardous ingredients, it will not contaminate the groundwater.

Protection board ranging in thickness from ⅜ inch (which is standard) to 2 inches can be put on to protect the membrane. See Fig. 3-48 A to H. In addition, plastic drain board may be used. The basic use for this waterproofing membrane is for vertical and horizontal building elements above and below grade such as walls, slabs, decks, and underground structures. Rub-R-Wall prevents the passage of water under hydrostatic, dynamic, or static pressure. When it is used on basement walls and floors inside the basement, it is possible to eliminate the threat from radon gas that plagues some areas of the United States. The material is made of rubber polymer and is applied in liquid form by spraying it onto walls and floors. The more complicated the surface to be waterproofed, the more reason to use a liquid applied membrane. The material can conform to all irregular shapes. It can be applied at low temperatures (15°F) provided substrates are

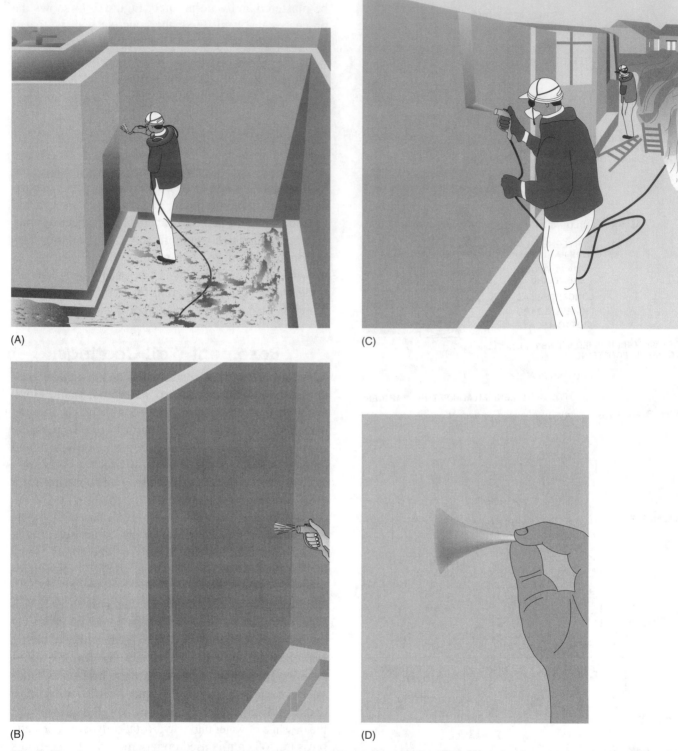

(A)

(B)

(C)

(D)

Fig. 3-48 *Green coated Rub-R-Wall is applied to basement walls before backfilling. (A) The first of three coats of the green polymer. (B) The second coat of three. (C) The third coat of the membrane being applied by spray gun. (D) The fresh membrane can already have much flexibility.*

dry and frost-free. Once the membrane is applied, it is impermeable to water. The excellent and tenacious bond of the membrane to substrates and protection course prevents the lateral movement of water between the membrane and substrate.

It is installed in a minimum of 40 mils thick, averaging 50 mils thick for waterproofing purposes. The only limitation is that the waterproofing must be protected from ultraviolet light rays (sun) and mechanical damage and should not be left permanently exposed.

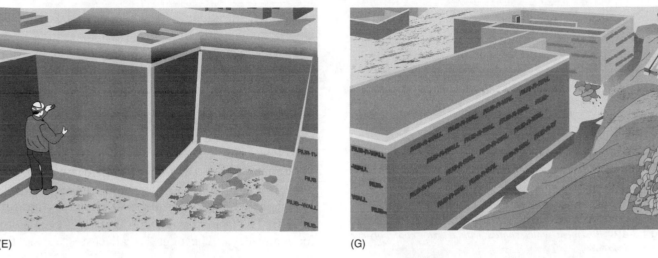

(F)

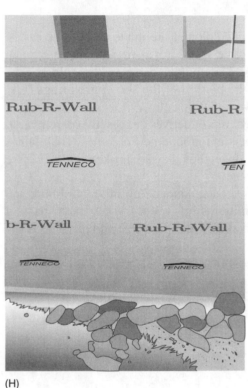

(H)

Fig. 3-48 *Continued. (E) Plastic flashing on corners and footer. (F) Flashing above the finished board. (G) Finished job with protective wall coating. (H) Closer look at the finished job.*(Rub-R-Wall)

However, do not backfill sooner than 24 hours after the membrane is applied. It will dry to the touch and not be sticky within 20 minutes. This coating does not require maintenance. Damaged areas can be easily repaired by spraying over affected areas. Cold joints or recoating is not a problem; newly applied material easily blends with existing membrane to provide a monolithic membrane.

Gray Wall

Gray wall is a 100% gray rubber coating that outperforms all asphalt-based dampproofing on the market today. It has 1400% elongation that allows the material to bridge small cracks in the foundation. It is applied at about one-half the thickness and amount of product as Rub-R-Wall. Protection board may also be used on this product. Gray wall offers seamless protection at an economical price. See Fig. 3-49.

TERMITES

Termite is the common name applied to white ants. They are neither all-white nor ants. In fact, they are closely related to cockroaches. Termites do, however, live in colonies somewhat in the same manner as ants do.

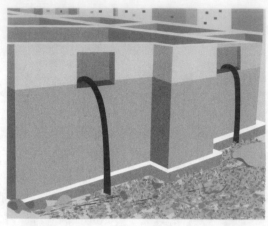

Fig. 3-49 *Basement walls coated with a gray coat polymer coating.* (Rub-R-Wall)

Most termite colonies are made up of three castes. The highest caste is the royal or reproductive group (Fig. 3-50A). The middle caste is the soldier. The worker is at the bottom of the social groupings (Fig. 3-50B). In every mature colony of termites, a group of young winged reproductives leaves the parent nest, mates, and sets out to found new colonies. Their wings are used only once; then they are broken off just before they seek a mate.

The worker caste is made up of small, blind, and wingless termites (Fig. 3-50B). They have pale or whitish soft bodies. Only their feet and heads are cov-

ered by a hard coating. The worker caste makes up the largest group within a colony.

The soldiers (Fig. 3-50C) have very long heads in proportion to their bodies and are responsible for protecting the colony against its enemies, usually ants. The soldiers are also blind and wingless. Because of this, the workers do all the work. They enlarge the nest, search for food and water, and make tunnels. They also take care of the soldiers since they have to be fed individually.

Termite castes contain both female and male. The kings live as long as the queens. The queens become as long as the males—3 inches—when they are full of eggs. They can lay many thousands of eggs a day for many days. The eggs for all castes appear the same.

Termites live in warm areas such as Africa, Australia, and the Amazon. They build nests as high as 20 feet, with the inside divided into chambers and galleries. They keep the king and queen separated in a closed cell. The workers carry the eggs away as fast as the queen lays them. The workers then care for the eggs until they hatch and then take care of the young until they grow up. Termites digest wood, paper, and other materials. They use cellulose for food. Most of the damage they do to homes and furniture is through tunneling. They have also been known to destroy books in search for cellulose to digest. They also do great damage to sugar cane and orange trees. They are considered a serious pest in many parts of the United States where they damage houses.

The best way to provide protection for a home is to follow some suggestions that have been developed through the years. There are about 2000 different species of termites known. About 40 species live in the United States and two species in Europe. They do not build large mounds in the United States or Europe, but do most of their damage out of sight.

Types of Termites

Three groups of termites exist in the United States. They are grouped according to their habits. The *subterranean* (underground) termites are the smallest and the most destructive, for they nest underground. They extend their habitat for long distances into wood structures. The *damp wood* termites live only in very moist wood. This type causes trouble only on the Pacific Coast. The *dry wood* termites need very little moisture. They are found to be destructive in the Southwest. The damp wood and dry wood types do not have a distinctive working caste (Fig. 3-51).

Once termites get into the house, they eat books, cloth, and furniture. They also attack bridges, trestles,

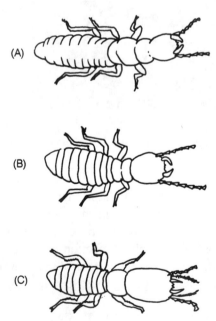

(A)

(B)

(C)

Fig. 3-50 *(A) The supplementary queen leaves the colony to mate, then sets up a new colony. She breaks off her wings before mating. (B) The worker termite is small and has a pale soft body. Workers gather food for the colony. (C) The soldier termite is extremely large when compared to the worker. The soldier has a hard head and defends the colony.*

Termite protection generally
not required, except that certain
localities may require protection
when a hazard exists.

Termite protection not required
(including Alaska).

Termite protection
required in all areas
(including Hawaii).

Termite protection
generally required.

Puerto Rico is an
area of severe termite
infestation. All lumber
should be pressure treated
per AWPB standards.

Fig. 3-51 *Note where the termites are located in the United States.*

and other wooden structures. They do more damage each year in the United States than fire.

Termite Protection

Most termites cannot live without water. The best approach is to eliminate their source of moisture. This can be done by applying chemicals to the soil around the foundation of a building or impregnating the wood with chemicals that repel or kill termites. Creosote and other types of chemicals are used around the footings of buildings and termite shields are used to keep them from reaching the wood sill (Fig. 3-52). Metal shields are also placed on copper water pipe or soil pipe to prevent their climbing up the metal pipes to reach the wood of the building (see Fig. 3-53). The builder must take every precaution to prevent infestation.

Infestation can be prevented by using minimum space between joists and the soil in the crawl space (see Fig. 3-54). Keep the space to a minimum of 18 inches. The minimum recommended space between girders and the soil is 18 inches. The lowest wood member of the exterior of a house should be placed 6 inches or more above the grade. Metal termite shields should be used on each side of a masonry wall (see Fig. 3-52A). Metal shields should be at least 24-gage galvanized iron. Concrete should be compacted as it is placed in the forms. This makes sure rock pockets and honeycombs are eliminated.

After a house is completed, make sure all wood scraps are removed. Any scraps buried in backfilling

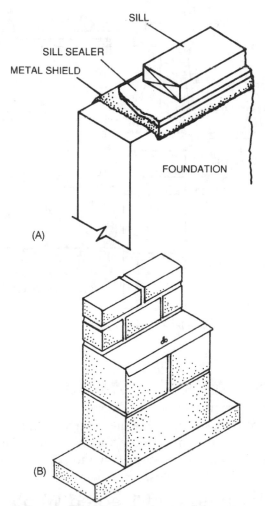

SILL

SILL SEALER

METAL SHIELD

FOUNDATION

(A)

(B)

Fig. 3-52 *(A) A metal shield is placed over the concrete foundation before the wood sill is applied. (B) Even masonry walls should have a termite shield in areas infested with termites.*

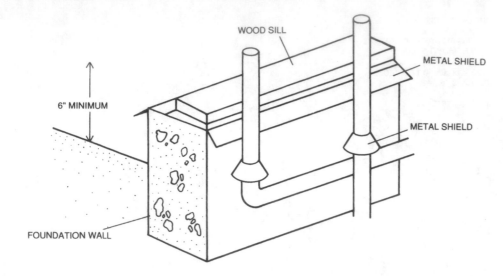

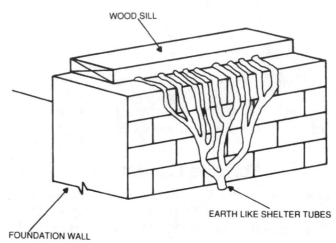

Fig. 3-53 *Note the placement of shield over pipes to protect the wooden floors from infestation.*

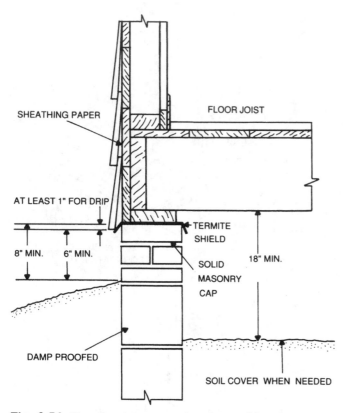

Fig. 3-54 *Note the minimum spacings for wood in reference to the grade.*

Fig. 3-55 *Termites are well known for their tunneling ability.*

ture (approximately 50 to 90°F), a source of oxygen, a moisture content above 20%, and a source of food (wood fiber). If any one of these conditions is removed, infestation will not occur. Chemical preservatives eliminate wood as a food source.

PRESSURE-TREATED WOOD

Treatment of wood adds to its versatility. Treatment with chemical preservatives protects wood that is exposed to the elements, is in contact with the ground, or is used in areas of high humidity. The treatment of wood allows you to build a project for outside exposure that will last a lifetime.

Properly treated, pressure-treated wood resists rot, decay, and termites and provides excellent service, even when exposed to severe conditions. It is wood that has been treated under pressure and under controlled conditions with chemical preservatives that penetrate deeply into the cellular structure.

or during grading become possible pockets for termite colonies. From there they can tunnel into a house (see Fig. 3-55). Make sure there are no fine cracks in the concrete foundation or loose mortar in the cement blocks. Termites have been known to build clay tunnels around metal shields.

Termites and Treated Wood

Decay and termite attack can occur when all four of the following conditions prevail: a favorable tempera-

Just because a wood is green in color doesn't mean it is pressure-treated wood. And just because it is pressure-treated doesn't mean there is enough of the chemical deep enough in the wood to prevent decay and keep insects at bay. Only in recent years has treated wood been readily available from local lumberyards.

In order to obtain the properly treated wood you should use the American Wood Preservers Bureau's recommendations. First, you should consider the wood. Not all wood is created equal. Most wood species don't readily accept chemical preservatives. To assist preservative penetration, the American Wood Preservers Association standards require incising for all species except southern pine, ponderosa pine, and red pine. Incising is a series of little slits along the grain of the wood which assists chemical penetration and uniform retention. Depth of penetration is important in providing a chemical barrier thick enough that any checking or splitting won't expose untreated wood to decay or insect attack.

To make sure the wood is strong enough for your intended use, you should always insist that the lumber you buy bears a lumber grade mark. See Fig. 3-56A. This is typical of the symbols (grade marks) found on southern pine lumber and plywood. See Fig. 3-56B.

Preservatives

There are three types of wood preservatives used. *Waterborne preservatives* are used in residential, commercial, recreational, marine, agricultural, and industrial applications. *Creosote* and mixtures of creosote and coal tar in heavy oil are used for railroad ties, utility poles, piles, and similar applications. *Pentachlorophenol* (Penta) in various solvents is used in industrial applications, utility poles, and some farm uses.

Waterborne preservatives are most commonly used in home construction due to their clean, odorless appearance. Wood treatment with waterborne preservatives can be stained or painted when dry. These preservatives also meet stringent EPA health guidelines. The preservative most commonly used in residential construction lumber is known by a variety of brand names, but in the trade it is known simply as CCA. The CCA stands for chromates copper arsenate. It also accounts for the green (copper) color in the treated wood.

CCA is a waterborne preservative. The chromium salts combine with the wood sugar to form an insoluble compound that renders the CCA preservative nonleachable. Some estimate the life of properly treated wood to exceed 100 years because the chemicals do not leach out in time or with exposure to the elements. However, in order to make sure the wood is properly treated after it has been cut, you must treat the exposed area with chemicals. It is a good idea to avoid field cuts and drilling in portions of treated wood that will be submerged in water.

Aboveground and In-Ground Treatment

The standard retentions most commonly found in lumberyards are 0.25 (pounds of chemical per cubic foot of wood) for aboveground use and 0.40 for below- (or

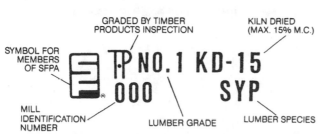

Fig. 3-56A *Two typical lumber grade marks.* (Southern Forest Products Association)

Southern Pine is the preferred lumber species for building the PermanentWood Foundation. Southern Pine lumber is readily available in a wide range of grades and sizes. In structural terms, it is one of the strongest softwoods.

Quality Southern Pine lumber is graded in accordance with the grading rules of the Southern Pine Inspection Bureau (SPIB).

Fig. 3-56B *Typical APA Grade Stamps for PWF construction.* (American Plywood-Association)

in-ground contact with) ground use. The only difference is a slightly higher concentration of chemical in 0.40 below-ground treatments. So if the aboveground (0.25) material is not available, below-ground (0.40) can be used instead. However, you should not use 0.25 material in contact with the ground.

Retention (lb/ft^3)	Use
0.25	Aboveground
0.40	Ground contact
0.60	Wood foundation
2.50	In saltwater

Only FDN (foundation) treatment should be used for wood foundation lumber and plywood applications. All pieces of lumber and plywood used for wood foundation applications must be identified by the FDN stamp (0.60 retention). That means the foundation should have a life longer than that of the rest of the structure.

Nails and Fasteners

Hot-dipped galvanized or stainless-steel nails and fasteners should be used to ensure maximum performance in treated wood. Such fasteners ensure permanence and prevent corrosion, which stains both the wood and its finishes. In structural applications, where a long service life is required, stainless steel, silicon, bronze, or copper fasteners are recommended. Smaller nails may be used with southern pine because of its greater nail-holding ability. Use 10d nails to fasten 2-inch dimensional lumber, 8d nails for fastening 1-inch boards to 2-inch dimensional lumber, and 6d nails for fastening 1-inch boards to 1-inch boards. The use of treated lumber adhesives can be considered for attaching deck boards to

Fig. 3-57 *The bark side of the lumber is exposed when using treated wood for a deck or other exposed surface.* (Southern Forest Products Association)

joists, in building fences, or in applications where the appearance of nail heads is not desired. For deck applications, fastening boards bark-side-up will help reduce surface checking and cupping. See Fig. 3-57.

Handling and Storing Treated Wood

When properly handled, pressure-treated wood is not believed to be a health hazard. Treated wood should be disposed of by ordinary trash collection or burial. It should not be burned. Prolonged inhalation of sawdust from untreated and treated wood should be avoided. Sawing should be performed outdoors while you are wearing a dust mask. Eye goggles should be worn when power-sawing or machining. Before eating, drinking, or using tobacco products, areas of skin that have come in contact with treated wood should be washed thoroughly. Clothes accumulating preservatives and sawdust should be laundered before reuse and washed separately from other household clothing.

Care should be taken to prevent splitting or excessively damaging the surface of the lumber, since this could permit decay organisms to get past the chemical barrier and start deterioration from within. Treated lumber should be stacked and stored in the same manner as untreated wood. Treated wood will also weather. If it is stored outside and exposed to the sun and elements, the green color will eventually turn to the characteristic gray, just as natural brown or redcolored wood does.

4
CHAPTER

Pouring Concrete Slabs & Floors

CONCRETE SURFACES ARE USED FOR MANY things. Slabs combine footings, foundations, and subfloors in one piece. Concrete floors are common in basements and in baths. Concrete is used outdoors to form stairs, driveways, patios, and sidewalks. Carpenters build the forms and, in some cases, also help pour and finish the concrete. After this chapter you should be able to

- Excavate
- Construct the forms
- Prepare the subsurface
- Lay drains and utilities
- Lay reinforcement
- Determine concrete needs
- Ensure correct pouring and surfacing

CONCRETE SLABS

Concrete slabs can combine footings, foundations, and subfloors as one piece. Slabs are cheaper to build than basements. In the past, basements were used as storage areas for furnaces, fuels, and ashes. Basements also held cooling and ventilation units and laundry areas. Today, these things are as easily built on the ground. However, in cold climates many people still prefer basements. They keep pipes from freezing and add warmth to the upper floors. They also provide storage space, play areas, and sometimes living areas.

Slabs are best used on level ground and in warm climates. They can be used where ground hardness is uneven. Slabs are good with split-level houses or houses on hills where the slab is used for the lower floor. See Fig. 4-1.

In the past, slabs did not make comfortable floors. They were very cold in the winter, and water easily condensed on them. Water could also seep up through

the slab from the ground. Today's building methods can solve most of these problems.

Most heat energy is lost at the edges of a slab. This loss is reduced by using rigid insulation. See Fig. 4-2. In extreme cases a warmer floor is needed. Heating ducts may be built into the slab flooring itself. This is an efficient method and provides an even temperature in the building. A warmer type of floor can also be built over the concrete floor. This is discussed later in this chapter.

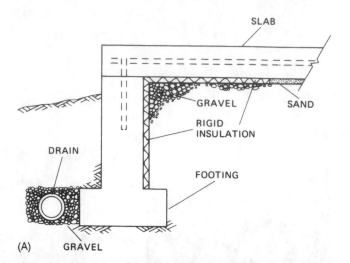

(A)

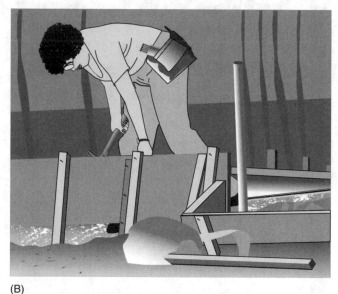

(B)

Fig. 4-2 (A) Insulation for a slab. (B) Forms for a slab. (Fox and Jacobs)

Sequence for Preparing a Slab

Slabs are used for many types of buildings and for several types of outdoor surfaces. In most cases, the general procedure is about the same. After the site is prepared and the building is located, this sequence is common:

1. Excavate.
2. Construct forms.

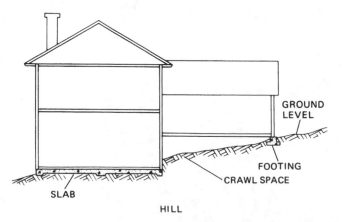

Fig. 4-1 A typical split-level home. The lower floor is a slab; the middle floor is a frame floor.

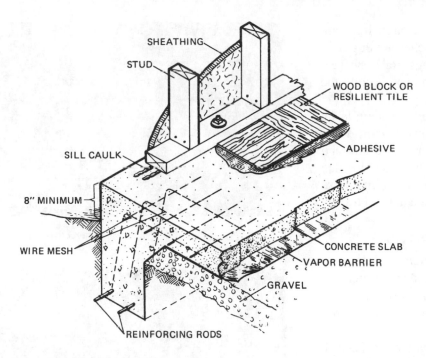

Fig. 4-3 *Combined slab and foundation (thickened-edge slab).*
(Forest Products Laboratory)

3. Prepare subsurface.
   a. Spread sand or gravel and level.
   b. Install drains, pipes, and utilities.
   c. Install moisture barrier.
4. Install reinforcement bar.
5. Construct special forms for lower levels, stairs, walks, etc.
6. Level tops of forms.
7. Determine concrete needs.
8. Pour concrete.
9. Tamp, level, and finish.
10. Embed anchors.
11. Cure and remove forms.

Types of Slabs

Slabs are built in two basic ways. Those called *monolithic* slabs are poured in one piece. They are used on level ground in warm climates.

In cold climates, however, the frostline penetrates deeper into the ground. This means that the footing must extend deeper into the ground to be below the frostline. Another method for slabs is often used where the footing is built separately. The two-piece slab is best for cold or wet climates. Figures 4-2 and 4-3 show both types of slabs.

Slab footings must rest beneath the frostline. This gives the slab stability in the soil. Slabs should be reinforced, but the amount of reinforcement needed varies. It depends on soil conditions and weights to be carried. On dry, stable soil, slabs need little rein-

forcement. Larger slabs or less stable soils need more reinforcement.

The top of the slab should be 8 inches aboveground. If the slab is above the rest of the ground, moisture under the slab can drain away from the building. The ground around the slab should also be sloped for the best drainage. See Fig. 4-4.

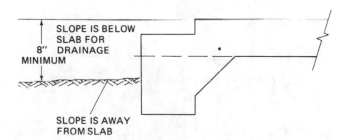

Fig. 4-4 *Drainage of a slab.*

Excavate

At this point, the building has already been located and batter boards are in place. Lines on the batter boards show the location of corners and walls.

Now, trenches or excavations are made for the footings, drains, and other floor features. These must all be deeper than the rest of the slab. Footings are, of course, around the outside edges. However, a slab big enough for a house should also have central footings.

The locating and digging for slab footings are the same as for foundation footings. There is no difference at all when a two-piece slab is to be made. The main difference for a monolithic slab lies in how the forms are made.

The trenches can now be dug. Rough lines are used for guides. As a rule, inner trenches are dug first. The trenches for the outside footings are done last. This is easier when machines are moved around the site.

The excavations are then checked for depth level. Remember from Chap. 3 that the lowest point determines the depth. Trenches are also dug for drains and sewer lines inside the slab. Then, trenches are dug from the slab to the main utility lines. Sewer lines must connect the slab to the main sewer line, and so forth.

Construct the Forms

After the excavation is done, the corners are relocated. Lines are restrung on batter boards and the corners are plumbed. See Fig. 3-16. The footing forms for two-piece slabs are made like standard footing or foundation forms. Refer to Chap. 3.

Lumber is brought to the place where it is needed, and then forms are constructed. See Fig. 4-5. Monolithic forms are made like footing forms. The top board is placed and leveled first. See Fig. 3-16. It is leveled with the corner first. Then its length is leveled and the ends are nailed to stakes. As before, double-headed nails should be used from the outside. The remaining form boards are then nailed in place.

Fig. 4-5 *Forms are erected.* (Fox and Jacobs)

Another method may be used for monolithic slabs with shallow footings. It is a very fast and inexpensive way. The form boards are put up before excavating. See Fig. 4-6. Be sure to carefully check the plans. Next, the sand or gravel is dumped inside the form area. See Fig. 4-7. The sand is spread evenly over the form area. As in Fig. 4-8, the outside footings are then dug by hand. Chalk lines are strung from the forms.

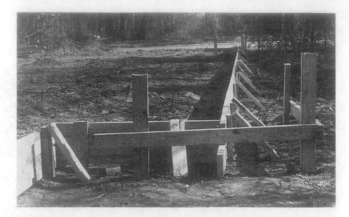

Fig. 4-6 *The forms are erected.*

Fig. 4-7 *Sand is dumped and spread in the form area.*

Fig. 4-8 *Excavations are made for outside footings. Plumbing is "roughed in."*

They are then used as guides to dig the central footings. See Fig. 4-9. Trenches for drains and sewers are then dug.

Fig. 4-9 *Excavations are also made for footings in the slab.*

Prepare the Subsurface

The ground under any type of slab must be prepared for moisture control. Water must be kept from seeping up through a slab. The water must also be drained from under the slab. This preparation is needed because water from rain and snow will seep under the slab.

Outside moisture can be reduced by using good siding methods. The edges of the slab can be stepped for brick. The sheathing can overlap the slab edge on other types of siding. As mentioned before, heat energy can be lost easily from slabs. The main area of loss is around the edges. Rigid foam insulation can be put under the slab's edges. This will reduce the heat energy losses of the slab.

Subsurface preparation After excavation, drains, water lines, and utilities are "roughed in." These should be placed for areas such as the kitchen, baths, and laundry. Water lines may be run in ceilings or beneath the slab. If they are to be beneath the slab, soft copper should be used. Extra length should be coiled loosely to allow for slab movement. Metal or plastic pipe may be used for the drains. Conduit (metal pipe for electrical wires) should be laid for any electrical wires to go under the slab. Wires should never be laid without the conduit. All openings in the pipes are then capped. See Fig. 4-10. This keeps dirt and concrete from clogging them.

The various pipes are then covered with sand or gravel. Sand or gravel is dumped in the slab area and carefully smoothed and leveled. Chalk lines are strung across the forms to check the level. See Fig. 4-11. It is also wise to install a clean-out plug between the slab and the sewer line. See Fig. 4-12.

Lay a vapor seal Once the sand is leveled, the moisture barrier is laid. The terms *vapor barrier, moisture barrier, vapor seal,* and *membrane* mean the same

Fig. 4-10 *All openings in pipes are covered. This prevents them from clogging with dirt or concrete.*

Fig. 4-11 *The sand is leveled. Chalk lines are used as guides.*

Fig. 4-12 *Trenches for drains and sewers are also excavated. Cleanout plugs are usually outside the slab.*

Fig. 4-13 *The vapor barrier is then laid. The reinforcement is also laid.*

thing. As a rule, plastic sheets are used for moisture barriers. The moisture barrier is laid so that it covers the whole subsurface area. To do this, several strips of material will be used. The strips should overlap at least 2 inches at the edges.

Insulate the edges Figure 4-2 shows insulation for a slab. Insulation is laid after placement of the vapor barrier. The insulation is placed around the outside edges. This is called *perimeter insulation*. The insulation should extend to the bottom of the footing. It should extend into the floor area at least 12 inches. A distance of 24 inches is recommended. Rigid foam at least 1 inch thick is used. Perimeter insulation is not always used in warm climates. However, the moisture barrier should always be used.

Reinforcement Reinforcement should always be used. The amount of reinforcement rods (rebar) used in the footing should conform to local codes. The slab should also be reinforced with mesh. This mesh is made of 10-gage wire. The wires are spaced 4 to 6 inches apart. Figure 4-13 shows the reinforced form ready for pouring. Where the soil is unstable, more reinforcement is needed. The amount is usually given on the plans.

As the rebar is laid, it is tied in place. The soft metal ties are twisted around the rebars. See Fig. 4-14. The mesh may also be held off the bottom by metal stakes. See Fig. 4-15. Mesh may also be lifted into place as the concrete is poured. Chairs hold rebar in place for pouring. See Fig. 4-16.

Fig. 4-14 *Rebar is tied together with soft wire "ties."*

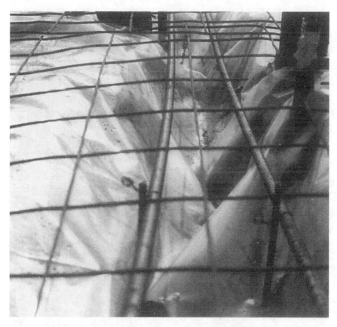

Fig. 4-15 *Short stakes hold mesh and rebar in place during pouring.*

Fig. 4-16 *Chairs hold rebar in place for pouring.* (Fox and Jacobs)

Special Shapes

Special shapes are used for several reasons. A stepped edge, as in Fig. 4-17, helps drainage. It prevents rainwater from flowing onto the floor surface. Lower sur-

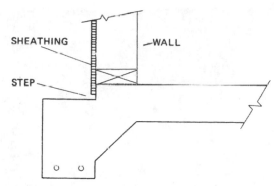

Fig. 4-17 *A stepped form aids drainage. The step also forms a base for brick siding.*

face areas are also common. They are used for garages, entryways, and so forth.

Step edges are easily formed. A 2 × 4 or 2 × 6 is nailed to the top of the form. See Fig. 4-18A and B. To lower a larger surface, an extension form is used. The first form is used as a nailing base. The extension form is built inside the outer form. The lower area can then be leveled separately. See Fig. 4-19.

(A)

(B)

Fig. 4-18 *A stepped edge is formed by nailing a board to the form.* (Fox and Jacobs)

Fig. 4-19 *A lower surface is used for a garage. (See top left through bottom right.)*

Pouring the Slab

First, the corners of the forms are leveled. See Fig. 4-20. A transit is used for large slabs, but small areas are leveled with a carpenter's level. See Chap. 2 for the leveling process with a transit.

Diagonals are checked for squareness, and all dimensions are checked.

Most concrete today is delivered to the building site. Usually, the concrete is not mixed by the carpenters. It is delivered by transit-mix trucks from a concrete company. The concrete is sold in units of cubic yards. Before the concrete can be ordered, the amount of concrete needed must be determined.

Estimating volume A formula is used to estimate the volume of concrete needed. The following formula is used:

$$\frac{L'}{3} \times \frac{W'}{3} \times \frac{T''}{36} = \text{cubic yards}$$

For example, a slab has a footing 1 foot wide. The slab is to be 30 feet wide and 48 feet long. The slab is to be 6 inches thick. From the formula, the amount required is

$$\frac{30}{3} \times \frac{48}{3} \times \frac{6}{36} = \frac{8640}{324}$$

$$\frac{10}{1} \times \frac{16}{1} \times \frac{1}{6} = \frac{160}{6}$$

$$= 26\frac{2}{3}$$

But the perimeter footings are 18 inches deep. They are not just 6 inches deep. Thus, a portion 12 inches

Fig. 4-20 *Corners of forms are leveled before pouring the slab.*
(Portland Cement)

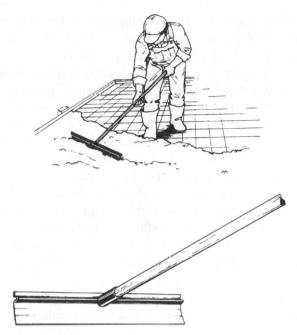

Fig. 4-21 *Concrete is spread evenly in the form.*

thick (18 minus 6) must be added. This additional amount of concrete is calculated as follows. The linear distance around the slab is

$$48 + 48 + 30 + 30 = 156$$

Then

$$\frac{L}{3} \times \frac{W}{3} \times \frac{T}{36} = \frac{1}{3} \times \frac{156}{3} \times \frac{12}{36}$$

$$= \frac{1}{3} \times \frac{52}{1} \times \frac{1}{3} = \frac{52}{9} = 5.7 \text{ cubic yards}$$

Thus, to fill the slab, the two elements are added:

$$\begin{array}{r} 26.7 \\ + 5.7 \\ \hline 32.4 \text{ cubic yards} \end{array}$$

Pouring To be ready, the carpenter sees to two things. First the forms must be done. Then the concrete truck must have a close access.

The builder must spread, carry, and level the concrete. The truck will only deliver it to the site. The truck driver can remain only a few minutes. The driver is not allowed to help work the concrete. As the concrete is poured, it should be spread and tamped. This is done with a board or shovel. See Fig. 4-21. The board or shovel is plunged into the concrete. Be careful not to cut the moisture barrier. Tamping helps get rid of air pockets. This makes the concrete solidly fill all the form.

After tamping, the concrete is leveled. This is done with a long board called a *strike-off*. See Fig. 4-22. The

(A)

(B)

Fig. 4-22 *(A) As the form is filled, it is leveled. This is done with a long board called a strike-off. (B) Using a "jitterbug" to remove trapped air from the concrete.* (Skrobarczk Properties)

ends rest across the top of the forms. The board is moved back and forth across the top. A short back-and-forth motion is used. If the board is not long enough, special supports are used. These are called *screeds*. A screed may be a board or pipe supported by metal pins. The screed is leveled with the tops of the forms. It is removed after the section of concrete is leveled. Any holes left by the screeds are then patched.

After leveling, the surface is treated using a "jitter-bug" to remove trapped air in the concrete. Then, it is floated. This is done after the concrete is stiff. However, the concrete must not have hardened. A finisher uses a float to tamp the surface gently. During tamping the float is moved across the surface. Large floats called *bull floats* are used. See Fig. 4-23. Floating lets the smaller concrete particles float to the top. The large particles settle. This gives a smooth surface to the concrete. Floats may be made of wood or metal.

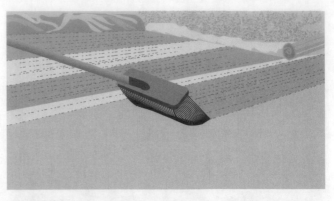

Fig. 4-24 *A broom can be used to make a lined surface.* (Portland Cement)

Fig. 4-25 *Hand trowels may be used to smooth small surfaces.* (Portland Cement)

Fig. 4-23 *After leveling, the surface is smoothed by "floating" with a bull float.* (Portland Cement)

After floating, the finish is done. A rough, lined surface can be produced. A broom is pulled across the top to make lines. This surface is easier to walk on in bad weather. See Fig. 4-24. For most flooring surfaces, a smooth surface is desired. A smooth surface is made by troweling. For small surfaces a hand trowel may be used, as in Fig. 4-25. However, for larger areas, a power trowel will be used. See Fig. 4-26.

Now the surface has been finished. Next, the anchors for the walls are embedded. Remember, the concrete is not yet hard. Do not let the anchors interfere with joist or stud spacing. The first anchor is embedded at about one-half the stud spacing. This would be 8 inches for 16-inch spacings. Anchors are placed at 4- to 8-foot intervals. Only two or three anchors are needed per wall.

Fig. 4-26 *A power trowel is used to finish large surfaces.* (Portland Cement)

The anchors are twisted deep into the concrete. The anchors are moved back and forth just a little. This settles the concrete around them. After the anchors are embedded, the surface is smoothed. A hand trowel is used. See Fig. 4-27.

The concrete is then allowed to harden and cure. Afterward, the form is removed. It takes about 3 days

Fig. 4-27 *Anchor bolts are embedded into foundation wall.*

to cure the concrete slab. During this time, work should not be done on the slab. When boards are used for the forms, they are saved. These boards are used later in the house frame. Lumber is not thrown away if it is still good. It is used where the concrete stains do not matter.

Expansion and Contraction

Concrete expands and contracts with heat and cold. To compensate for this movement, expansion joints are needed. Expansion joints are used between sections. The expansion joint is made with wood, plastic, or fiber. Joint pieces are placed before pouring. Such pieces are used between foundations and basement walls or between driveways and slabs. See Fig. 4-28. Often, the screed is made of wood. This can be left for the expansion joint.

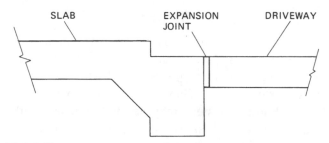

Fig. 4-28 *Expansion joints are used between large, separate pieces.*

Joints

Other joints are also used to control cracking. These, however, are shallow grooves troweled or cut into the concrete. They may be troweled in the concrete as it hardens. The joints may also be cut with a special saw blade after the concrete is hard. The joints help offset

and control cracking. Concrete is very strong against compression. However, it has little strength against bending or twisting. Note, for example, how concrete sidewalks break. The builder only tries to control breaks. They cannot be prevented. The joints form a weak place in concrete. The concrete will break at the weak point. But the crack will not show in the joint. The joint gives a better appearance when the concrete cracks. Figure 4-29 shows a troweled joint. Figure 4-30 shows a cut joint.

Fig. 4-29 *Joints may be troweled into a surface.* (Portland Cement)

Fig. 4-30 *Some joints may be cut with saws.*

CONCRETE FLOORS

Concrete is used for floors in basements and commercial buildings. The footings and foundations are built before the floor is made. An entire house may be built before the basement floor is made. See Fig. 4-31. The concrete can be poured through windows or floor openings.

Fig. 4-31 *Footings in a basement later become part of the basement floor.*

A concrete floor is made like a slab. It is also the way to make the two-piece slab. First, the ground is prepared. Drains, pipes, and utilities are positioned and covered with gravel. Next, coarse sand or gravel is leveled and packed. A plastic-film moisture barrier is placed to reach above the floor level. Perimeter insulation is laid as indicated. See Fig. 4-32. Reinforcement is placed and tied. Rigid foam insulation can also be used as an expansion joint. Finally, the floor is poured, tamped, leveled, and finished. Separate footings should be used to support beams and girders. Figure 4-33 shows this procedure.

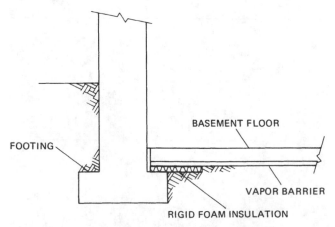

Fig. 4-32 *Rigid foam should be laid at the edges of the basement floor.*

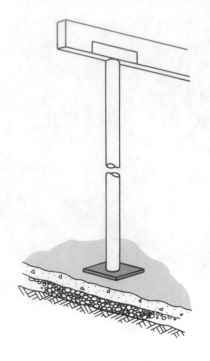

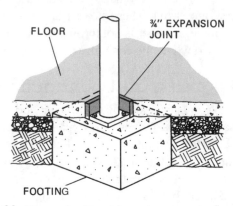

Fig. 4-33 *Floors should not be major supports. Separate footings should be used.*

Stairs

Entrances may be made as a part of the main slab. However, stairs used with slabs are poured separately. Separate stairs are used with slabs as an access in steep areas. They are also used with steep lawns. Forms for steps may be made by two methods. The first method uses short parallel boards. See Fig. 4-34. The second method uses a long stringer. Support boards are added as in Fig. 4-35. The top of the stair tread is left open in both types of forms. This lets the concrete step be finished. The bottoms of the riser forms are beveled. This lets a trowel finish the full surface. The stock used for building supports should be 2-inch lumber. This keeps the weight of the concrete from bulging the forms.

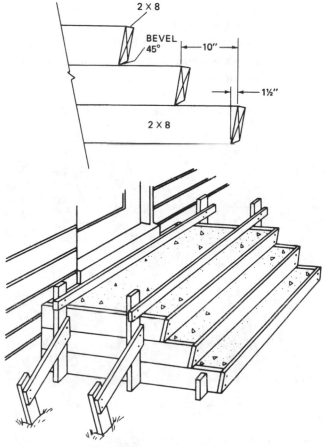

Fig. 4-34 *A form for steps.*

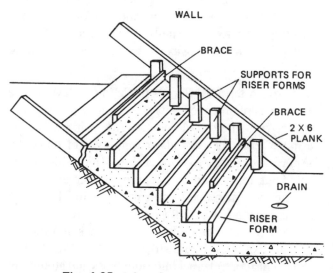

Fig. 4-35 *A form for steps against a wall.*

SIDEWALKS AND DRIVEWAYS

Sidewalks and driveways should not be a part of a slab. They must float free of the building. Where they touch, an expansion joint is needed. Sidewalks and driveways should slope away from the building. This lets water drain away from the building. Then, special methods are used for drainage. Often, a ledge about 1½ inches high

will be used. The floor of the building is then higher than the driveway. One side of the driveway is also raised. Water will then stop at the ledge and will flow off to the side. This keeps water from draining into the building.

Two-inch-thick lumber should be used for the forms. It should be 4 or 6 inches wide. The width determines slab thickness. Commercial forms may also be used. See Fig. 4-36.

Fig. 4-36 *Special forms may be used for sidewalks and driveways.* (Proctor Products)

Sidewalks

Sidewalks are usually 3 feet wide and 4 inches thick. Main walks or entryways may be 4 feet or wider. No reinforcement is needed for sidewalks on firm ground, though it may be used for sidewalks on soft ground. A sand or gravel fill is used for support of sidewalks on wet ground. The earth and fill should be tamped solid. Figure 4-37 shows a sidewalk form. The slab is poured, leveled, and finished like other surfaces.

Fig. 4-37 *Sidewalk forms are made with 2×4 lumber.*

Driveways

Driveways should be constructed to handle great weight. Reinforcement rods or mesh should be used in driveways. Slabs 4 inches thick may be used for passenger cars. Six-inch-thick slabs are used where trucks are expected. The standard driveway is 12 feet wide. Double driveways are 20 feet wide. Other dimensions are shown in Fig. 4-38. The slab is poured, leveled, and finished like other surfaces.

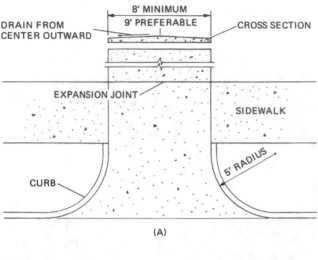

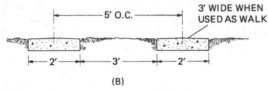

Fig. 4-38 *Driveway details: (A) Single-slab driveway. (B) Ribbon-type driveway.* (Forest Products Laboratory)

SPECIAL FINISHES AND SURFACES

Concrete may be finished in several ways. Different surface textures are used for better footing or appearance. Also, concrete may be colored. It may also be combined with other materials.

Different Surface Textures

Different surface textures may be used for appearance. However, the most common purpose is for better footing and tire traction, especially in bad weather. Several methods may be used to texture the surface. The surface may be simply floated. This gives a smooth, slightly roughened surface. Floated surfaces are often used on sidewalks.

Brushing The surface may also be brushed. Brushing is done with a broom or special texture brush. The pattern may be straight or curved. For a straight pattern, the brush is pulled across the entire surface. Refer again to Fig. 4-24. For swirls, the brush can be moved in circles.

Pebble finish Pebbles can be put in concrete for a special appearance. See Fig. 4-39. The pebble finish is

not difficult to do. As the concrete stiffens, pebbles are poured on the surface. The pebbles are then tamped into the top of the concrete. The pebbles in the surface are leveled with a board or a float. Some hours later, the fine concrete particles may be hosed away.

Fig. 4-39 *A pebbled concrete surface.*

Color additives Color may be added to concrete for better appearance. The color is added as the concrete is mixed and is uniform throughout the concrete. The colors used most are red, green, and black. The surface of colored concrete is usually troweled smooth. Frequently, the surface is waxed and polished for indoor use.

Terrazzo A terrazzo floor is made in two layers. The first layer is plain concrete. The second layer is a special type of white or colored concrete. Chips of stone are included in the second, or top, layer of concrete. The top layer is usually about ½ to 1 inch thick. It is leveled but not floated. The topping is then allowed to set. After it is hardened, the surface is finished. Terrazzo is finished by grinding it with a power machine. This grinds the surface of both the stones and the cement until smooth. Metal strips are placed in the terrazzo to help control cracking. Both brass and aluminum strips are used. Figure 4-40 shows this effect. The result is a durable finish with natural beauty. The floor may be waxed and polished.

Terrazzo will withstand heavy foot traffic with little wear. It is often used in buildings such as post offices or schools. It is also easy to maintain.

Ceramic tile, brick, and stone Concrete may also be combined with ceramic tile or brick. The result is a better-appearing floor. The floor contrasts the concrete

Fig. 4-40 *A terrazzo floor is smooth and hard. It can be used in schools and public buildings.* (National Terrazzo and Mosaic Association)

Fig. 4-41 *A concrete floor is being placed over a plywood floor. The concrete will make the floor more durable.*

and the brick or tile. Also, the concrete may be finished in several ways. This gives more variety to the contrast. For example, a pebble finish on the concrete may be used with special brick. Bricks are available in a variety of shapes, colors, and textures.

Ceramic tile comes in many sizes. The largest is 12 × 12 inches. Several shapes are also used. Glazed tile is used in bathrooms because the glaze seals water from the tile. Unglazed tile also has many uses, but it is not waterproof.

To set tile, stone, or bricks in a concrete floor, you must have a lowered area. The pieces are laid in the lowered area. The area between the pieces is filled with concrete, grout, or mortar.

Stone, tile, and brick are used. They add contrast and beauty and resist wear. They are easy to clean and resist oil, water, and chemicals.

Concrete over wood floors Concrete is also used for a surface over wood flooring. Figure 4-41 shows this kind of floor being made. A concrete topping on a floor has several advantages. These floors are harder and more durable than wood. They resist water and chemicals and may be used in hallways and rest rooms.

Wood over concrete Wood may be used for a finish floor over a concrete floor. The wooden floor is warmer in cold climates. Because it does not absorb heat as does concrete, energy can be saved. The use of wood can also improve the appearance of the floor.

Two methods are used for putting wood over a concrete floor. The first method is the older. A special glue, called *mastic*, is spread over the concrete. Then strips of wood are laid on the mastic. A wooden floor may be nailed to these strips. Figure 4-42A and B shows a cross section.

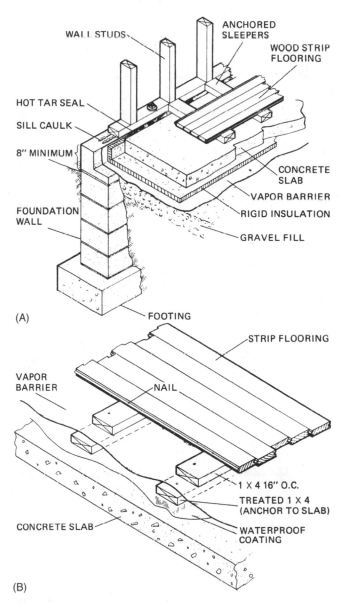

Fig. 4-42 *(A) Base for wood flooring on a concrete slab with vapor barrier under slab. (B) Base for wood flooring on a concrete slab with no vapor barrier under slab.* (Forest Products Laboratory)

For better energy savings, a newer method is used. In this method, rigid insulation is laid on the concrete. Mastic may be used to hold the insulation to the floor. See Fig. 4-43. The wooden floor is laid over the rigid insulation. Plywood or chipboard underlayment can also be used. The floor may then be carpeted. If desired, special wood surfaces can be used.

ENERGY FACTORS

There are two ways of saving energy that are used with concrete floors. The first is to insulate around the edges of the slab. The second is to cover the floor with insulated material. Both methods have been mentioned previously.

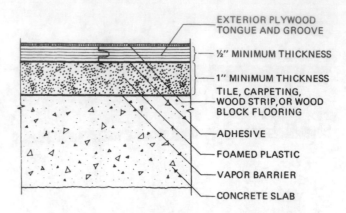

EXTERIOR PLYWOOD
TONGUE AND GROOVE
½" MINIMUM THICKNESS
1" MINIMUM THICKNESS
TILE, CARPETING,
WOOD STRIP, OR WOOD
BLOCK FLOORING
ADHESIVE
FOAMED PLASTIC
VAPOR BARRIER
CONCRETE SLAB

NOTE: Vapor barrier may be omitted if an effective vapor barrier is in place beneath the slab.

Fig. 4-43 *A wood floor may be laid over insulation. The insulation is glued to the concrete floor.* (American Plywood Association)

5
CHAPTER

Building
Floor
Frames

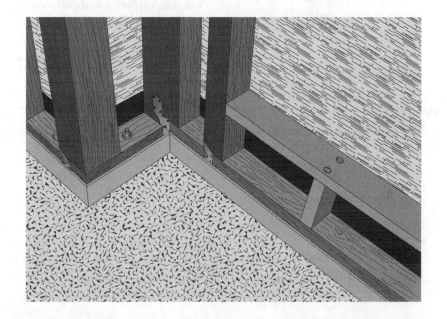

IN THIS UNIT YOU WILL LEARN HOW TO BUILD floor frames. You will learn how to make floors over basements and crawl spaces. You will also learn how to make openings for stairs and other things. Overall, you will learn how to

- Connect the floor to the foundation
- Place needed girders and supports
- Lay out the joist spacings
- Measure and cut the parts
- Put the floor frame together
- Lay the subflooring
- Build special framing
- Alter a standard floor frame to save energy

INTRODUCTION

Floors form the base for the rest of the building. Floor frames are built over basements and crawl spaces. Houses built on concrete slabs do not have floor frames. However, multilevel buildings may have both slabs and floor frames.

First the foundation is laid. Then the floor frame is made of posts, beams, sill plates, joists, and a subfloor. When these are put together, they form a level platform. The rest of the building is held up by this platform. The first wooden parts are called the *sill plates*. The sill plates are laid on the edges of the foundations. Often, additional supports are needed in the middle of the foundation area. See Fig. 5-1. These are called *midfloor supports* and may take several forms. These supports may be made of concrete or masonry. Wooden posts and metal columns are also used. Wooden timbers called *girders* are laid across the central supports. Floor joists then reach (span) from the sill on the foundation to the central girder. The floor

joists support the floor surface. The joists are supported by the sill and girder. These in turn rest on the foundation.

Two types of floor framing are used on multistory buildings. The most common is the platform type. In platform construction each floor is built separately. The other type is called the balloon frame. In balloon frames the wall studs reach from the sill to the top of the second floor. Floor frames are attached to the long wall studs. The two differ on how the wall and floor frames are connected. These will be covered in detail later.

SEQUENCE

The carpenter should build the floor frame in this sequence:

1. Check the level of foundation and supports.
2. Lay sill seals, termite shields, etc.
3. Lay the sill.
4. Lay girders.
5. Select joist style and spacing.
6. Lay out joists for openings and partitions.
7. Cut joists to length and shape.
8. Set joists.
   a. Lay in place.
   b. Nail opening frame.
   c. Nail regular joists.
9. Cut scabs, trim joist edges.
10. Nail bridging at tops.
11. Lay subfloor.
12. Nail bridging at bottom.
13. Trim floor at ends and edges.
14. Cut special openings in floor.

SILL PLACEMENT

The sill is the first wooden part attached to the foundation. However, other things must be done before the sill is laid. When the anchors and foundation surface are adequate, a seal must be placed on the foundation. The seal may be a roll of insulation material or caulking. If a metal termite shield is used, it is placed over the seal. Next, the sill is prepared and fitted over the anchor bolts onto the foundation. See Fig. 5-2.

The seal forms a barrier to moisture and insects. Roll-insulation-type material, as in Fig. 5-3, may be used. The roll should be laid in one continuous strip with no joints. At corners, the rolls should overlap about 2 inches.

Fig. 5-1 *The piers and foundation walls will help support the floor frame.*

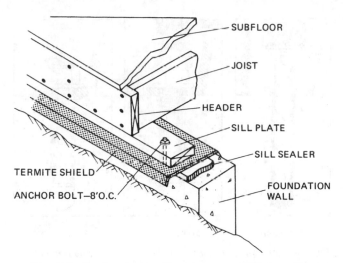

Fig. 5-2 *Section showing floor, joists, and sill placement.*

To protect against termites, two things are often done. A solid masonry top is used. Metal shields are also used. Some foundation walls are built of brick or concrete blocks that have hollow spaces. They are sealed with mortar or concrete on the top. A solid concrete foundation provides the best protection from termite penetration.

However, termites can penetrate cracks in masonry. Termites can enter a crack as small as ¹⁄₆₄ inch in width. Metal termite shields are used in many parts of the country. Figure 5-4 shows a termite shield installed.

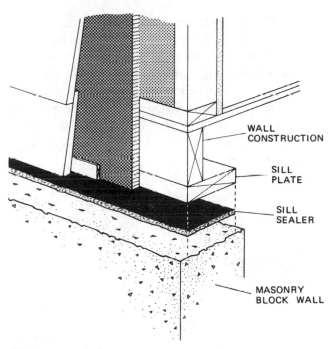

Fig. 5-3 *A seal fills in between the top of the foundation wall and the sill. It helps conserve energy by making the sill more weathertight.* (Conwed)

Anchor the Sill

The sill must be anchored to the foundation. The anchors keep the frame from sliding from the foundation. They also keep the building from lifting in high winds. Three methods are used to anchor the sill to the foun-

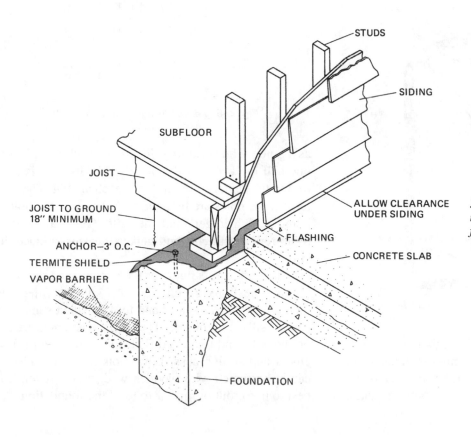

Fig. 5-4 *Metal termite shield used to protect wood over foundation.* (Forest Products Laboratory)

dation. The first uses bolts embedded in the foundation, as in Fig. 5-5. Sill straps are also used and so are
special drilled bolt anchors. See Figs. 5-6 and 5-7.
Special masonry nails are also sometimes used; however, they are not recommended for anchoring exterior
walls. It is not necessary to use many anchors per wall.
Anchors should be used about every 4 feet, depending
upon local codes. Anchors may not be required on
walls shorter than 4 feet.

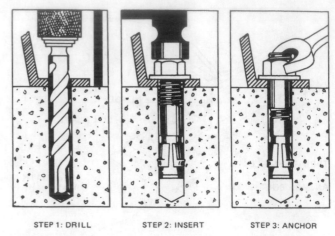

Fig. 5-7 *Anchor holes may be drilled after the concrete has set.*
(Hilti-Fastening Systems)

STEP 1: DRILL   STEP 2: INSERT   STEP 3: ANCHOR

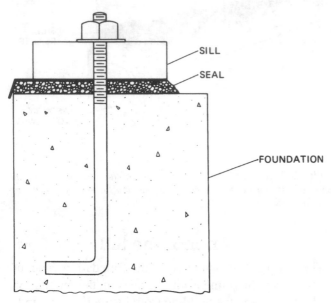

Fig. 5-5 *Anchor bolt in foundation.*

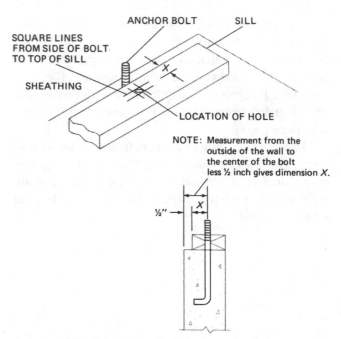

Fig. 5-8 *Locate the holes for anchor bolts.*

Fig. 5-6 *Anchor straps or clips can be used to anchor the sill.*

The anchor bolts must fit through the sill. The
holes are located first. Washers and nuts are taken from
the anchor bolts. The sill board is laid next to the bolts.
See Fig. 5-8. Lines are marked using a framing square

as a guide. The sheathing thickness is subtracted from
one-half the width of the board. This distance is used
to find the center of the hole for each anchor. The centers for the holes are then marked. As a rule the hole is
bored ¼ inch larger than the bolt. This leaves some
room for adjustments and makes it easier to place the
sill.

Next, the sill is put over the anchors and the spacing and locations are checked. All sills are fitted and
then removed. Sill sealer and termite shields are laid,
and the sills are replaced. The washers and nuts are put
on the bolts and tightened. The sill is checked for levelness and straightness. Low spots in the foundation
can be shimmed with wooden wedges. However, it is
best to use grout or mortar to level the foundation.

Special masonry nails may be used to anchor interior walls on slabs. These are driven by sledge hammers or by nail guns. The nail mainly prevents side slippage of the wall. Figure 5-9 shows a nail gun application.

Fig. 5-9 *A nail gun can be used to drive nails in the slab and for toenailing.* (Duo-Fast)

Setting Girders

Girders support the joists on one end. Usually the girder is placed halfway between the outside walls. The distance between the supports is called the *span*. The span on most houses is too great for joists to reach from wall to wall. Central support is given by girders.

Determine girder location Plans give the general spacing for supports and girders. Spans up to 14 feet are common for 2- × 10-inch or 2- × 12-inch lumber. The girder is laid across the leveled girder supports. A chalk line may be used to check the level. The support may be shimmed with mortar, grout, or wooden wedges. The supports are placed to equalize the span. They also help lower expense. The piers shown in Fig. 5-1 must be leveled for the floor frame.

The girder is often built by nailing boards together. Figure 5-10 shows a built-up girder. Girders are often made of either 2- × 10-inch or 2- × 12-inch lumber. Joints in the girder are staggered. The size of the girder and joists is also given on the plans.

There are several advantages to using built-up girders. First, thin boards are less expensive than thick ones. The lumber is more stable because it is drier. There is less shrinkage and movement of this type of girder. Wooden girders are also more fire-resistant than steel girders. Solid or laminated wooden girders take a long time to burn through. They do not sag or break until they have burned nearly through. Steel, on the other hand, will sag when it gets hot. It only takes a few minutes for steel to get hot enough to sag.

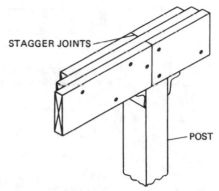

Fig. 5-10 *Built-up wood girder.* (Forest Products Laboratory)

The ends of girders can be supported in several ways. Figure 5-11A and B shows two methods. The ends of girders set in walls should be cut at an angle. See Fig. 5-12. In a fire, the beam may fall free. If the ends are cut at an angle, they will not break the wall.

Metal girders should have a wooden sill plate on top. This board forms a nail base for the joists. Basement girders are often supported by post jacks. See Fig. 5-13. Post jacks are used until the basement floor is finished. A support post, called a *lolly column*, may be built beneath the girder. It is usually made from 2 × 4 lumber. Walls may be built beneath the girder. In many areas this is done so that the basement may later be finished out as rooms.

JOISTS

Joists are the supports under the floor. They span from the sill to the girder. The subfloor is laid on the joists.

Lay Out the Joists

Joists are built in two basic ways. The first is the *platform method*. The platform method is the more common method today. The other method is called *balloon framing*. It is used for two-story buildings in some areas. However, the platform method is more common for multistory buildings.

Joist spacing The most common spacing is 16 inches. This makes a strong floor support. It also allows the carpenter to use standard sizes. However, 12-inch and 24-inch spacings are also used. The spacing depends upon the weight the floor must carry. Weight comes from people, furniture, and snow, wind, and rain. Local building codes will often tell what joist spacings should be used.

Joist spacings are given by the distance from the center of one board to the center of the next. This is called the distance *on centers*. For a 16-inch spacing, it would be written *16 inches O.C.*

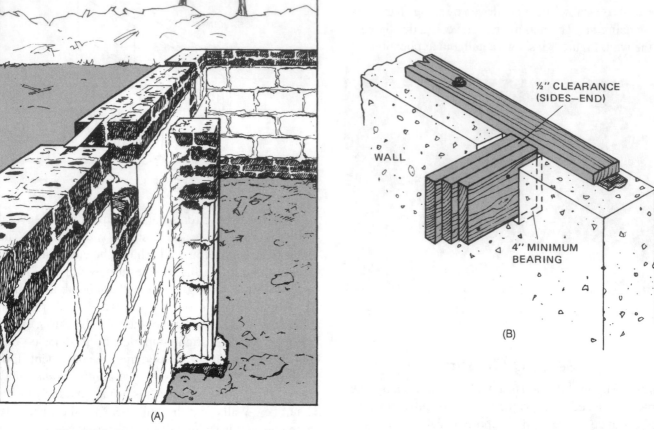
(A)

(B)

Fig. 5-11 *Two methods of supporting girder ends: (A) Projecting post. (B) Recessed pocket.* (Forest Products Laboratory)

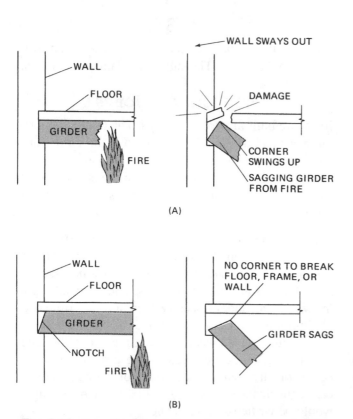
(A)

(B)

Fig. 5-12 *For solid walls, girder ends must be cut at an angle.*

Fig. 5-13 *A post jack supports the girder until a column or a wall is built.*

Modular spacings are 12, 16, or 24 inches O.C. These modules allow the carpenter to use standard-size sheet materials easily. The standard-size sheet is 48 × 96 inches (4 × 8 feet). Any of the modular sizes divides evenly into the standard sheet size. By using modules, the amount of cutting and fitting is greatly reduced. This is important since sheet materials are used on subfloors, floors, outside walls, inside walls, roof decks, and ceilings.

Joist layout for platform frames The position of the floor joists may be marked on a board called a *header*. The header fits across the end of the joists. See Fig. 5-14.

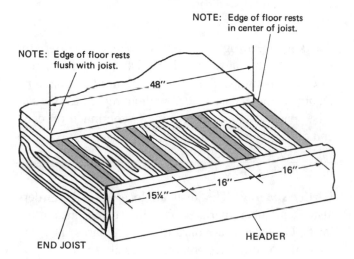

Fig. 5-14 *The positions of the floor joists may be marked on the header.*

Joist spacing is given by the distance between centers. However, the center of the board is a hard mark to use. It is much easier to mark the edge of a board. After all, if the centers are spaced right, the edges will be too!

The header is laid flat on the foundation. The end of the header is even with the end joist. The distance from the end of the header to the edge of the first joist is marked. However, this distance is not the same as the O.C. spacing. See Fig. 5-15. The first distance is always ¾ inch less than the spacing. This lets the edge of the flooring rest flush, or even, with the outside edge of the joist on the outside wall. This will make laying the flooring quicker and easier.

The rest of the marks are made at the regular O.C. spacing. See Fig. 5-15. As shown, an X indicates on which side of the line to put the joist.

Mark a pole It is faster to transfer marks than to measure each one. The spacing can be laid out on a board first. The board can then be used to transfer spacings. This board is called a *pole*. A pole saves time because measurements are done only once. To transfer the marks, the pole is laid next to a header. Use a square to

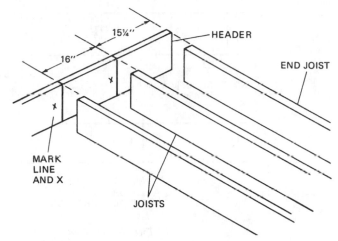

Fig. 5-15 *The first joist must be spaced ¾ inch less than the O.C. spacing used. Mark from end of header.*

project the spacing from the pole to the header. A square may be used to check the "square" of the line and mark.

Joists under walls Joists under walls are doubled. There are two ways of building a double joist. When the joist supports a wall, the two joists are nailed together. See Fig. 5-16. Pipes or vents sometimes go through the floors and walls. Then, a different method is used. See Fig. 5-17. The joists are spaced approximately 4 inches

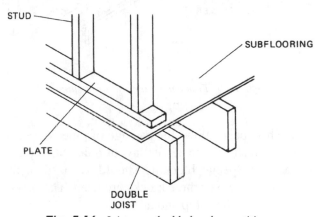

Fig. 5-16 *Joists are doubled under partitions.*

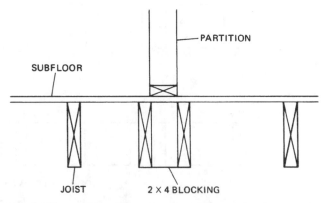

Fig. 5-17 *Double joists under a partition are spaced apart when pipes must go between them.*

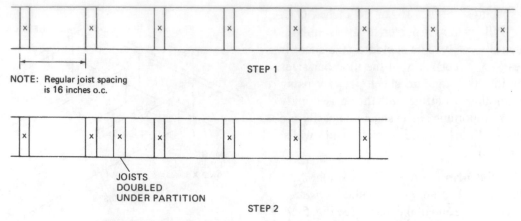

NOTE: Regular joist spacing is 16 inches o.c.

STEP 1

JOISTS
DOUBLED
UNDER PARTITION

STEP 2

Fig. 5-18 *Header joist layout. Next, add the joists for partitions.*

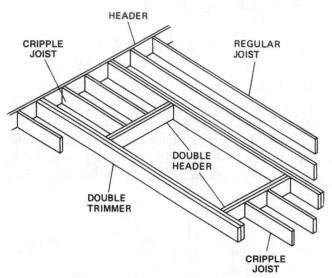

Fig. 5-19 *Frame parts for a floor opening.*

apart. This space allows the passage of pipes or vents. Figure 5-18 shows the header layout pole with a partition added. Special blocking should be used in the double joist. Two or three blocks are used. The blocking serves as a fire-stop and as bracing.

Joists for openings Openings are made in floors for stairs and chimneys. Double joists are used on the sides of openings. They are called *double trimmers*. Double trimmers are placed without regard for regular joist spacings. Regular spacing is continued on each end of the opening. Short "cripple" joists are used. See Fig. 5-19. A pole can show the spacing for the openings. See Fig. 5-20.

Girder spacing The joists are located on the girders also. Remember, marks do not show centerlines of the joists. Centerlines are hard to use, so marks show the edge of a board. These marks are easily seen.

Balloon layout Balloon framing is different. See Fig. 5-21. The wall studs rest on the sill. The joists and the studs are nailed together as shown. However, the end joists are nailed to the end wall studs.

The first joist is located back from the edge. The distance is the same as the wall thickness. The second joist is located by the first wall stud.

The wall stud is located first. The first edge of the stud is ¾ inch less than the O.C. spacing. For 16 inches O.C., the stud is 15¼ inches from the end. A 2-inch

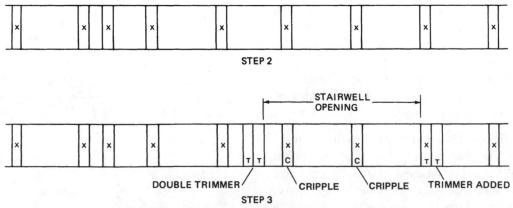

STEP 2

STAIRWELL
OPENING

DOUBLE TRIMMER   CRIPPLE   CRIPPLE   TRIMMER ADDED

STEP 3

Fig. 5-20 *Add the trimmers for the opening to the layout pole.*

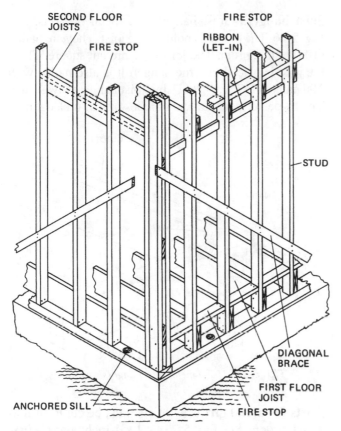

Fig. 5-21 *Joist and stud framing used in balloon construction.*
(Forest Products Laboratory)

stud will be 1½ inches thick. Thus, the edge of the first joist will be 16¾ inches from the edge.

Cut Joists

The joists span, or reach, from the sill to the girder. Note that joists do not cover the full width of the sill. Space is left on the sill for the joist header. See Fig. 5-4. For lumber 2 inches thick, the spacing would be 1½ inches. Joists are cut so that they rest on the girder. Four inches of the joist should rest on the girder.

The quickest way to cut joists is to cut each end square. Figure 5-22 shows square-cut ends. The joists

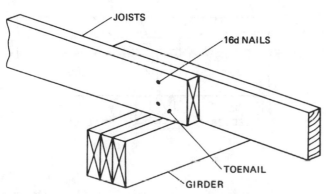

Fig. 5-22 *Joists may overlap on the girder. The overlap may be long or short.*

overlap across the girder. The ends rest on the sill with room for the joist header. This way the header fits even with the edge.

It is sometimes easier to put the joist header on after the joists are toenailed and spaced. Or, the joist header may be put down first. The joists are then just butted next to the header. But it is very important to carefully check the spacing of the joists before they are nailed to either the sill or the header.

Other Ways to Cut Joists

Ends of joists may be cut in other ways. The ends may be aligned and joined. Ends are cut square for some systems. For others, the ends are notched. Metal girders are sometimes used. Then, joists are cut to rest on metal girders.

End-joined joists The ends of the joists are cut square to fit together. The ends are then butted together as in Fig. 5-23. A gusset is nailed (10d) on each side to hold the joists together. Gussets may be made of either plywood or metal. This method saves lumber. Builders use it when they make several houses at one time.

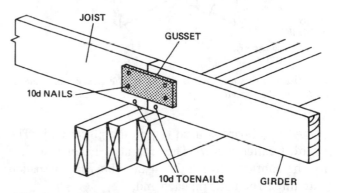

Fig. 5-23 *Joists may be butt-joined on the girder. Gussets may be used to hold them. Plywood subfloor may also hold the joists.*

Notched and lapped joists Girders may be notched and lapped. See Fig. 5-24. This connection has more interlocking but takes longer and costs more. First, the notch is cut on the end of the joists. Next a 2 × 4-inch joist support is nailed (16d) on the girder. Nails should be staggered 6 or 9 inches apart. The joists are then laid in place. The ends overlap across the girder.

Joist-girder butts This is a quick method. With it the top of the joist can be even with the top of the girder.

A 2 × 4-inch ledger is nailed to the girder with 16d common nails. See Fig. 5-25. The joist rests on the ledger and not on the girder. This method is not as strong.

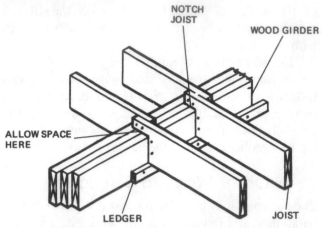

Fig. 5-24 *Joists can be notched and lapped on the girder.* (Forest Products Laboratory)

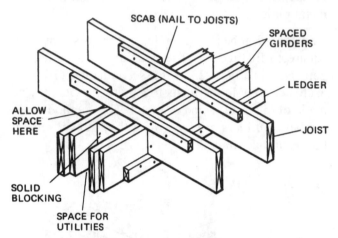

Fig. 5-25 *Joists may be butted against the girder. Note that the girder may be spaced for pipes, etc.* (Forest Products Laboratory)

Also, a board is used to join the girder ends. The board is called a *scab*. The scab also makes a surface for the subfloor. It is a 2 × 4-inch board. It is nailed with three 16d nails on each end.

Joist hangers Joist hangers are metal brackets. See Fig. 5-26. The brackets hold up the joist. They are nailed (10d) to the girder. The joist ends are cut square. Then, the joist is placed into the hanger. It is also nailed with 10d nails as in Fig. 5-26. Using joist hangers saves time. The carpenter need not cut notches or nail up ledgers.

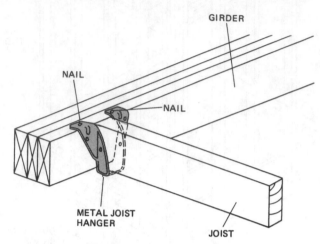

Fig. 5-26 *Using metal joist hangers saves time.*

Joists for metal girders Joists must be cut to fit into metal girders. See Fig. 5-27. A 2 × 4-inch board is first bolted to the metal girder. The ends of the joist are then beveled. This lets the joist fit into the metal girder. The joist rests on the board. The board also is a nail base for the joist. The tops of the joists must be scabbed. The scabs are made of 2 × 4-inch boards. Three 16d nails are driven into each end.

Setting the Joists

Two jobs are involved in setting joists. The first is laying the joists in place. The second is nailing the joists. The carpenter should follow a given sequence.

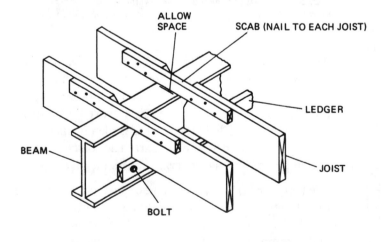

(A)

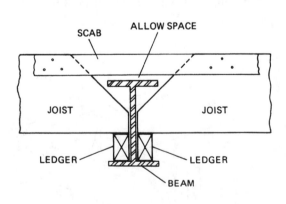

(B)

Fig. 5-27 *Systems for joining joists to metal girders.* (Forest Products Laboratory)

Lay the joists in place First, the header is toenailed in place. Then the full-length joists are cut. Then they are laid by the marks on the sill or header. Each side of the joist is then toenailed (10d) to the sill. See Fig. 5-28. Joists next to openings are not nailed. Next, the ends of the joists are toenailed (10d) on the girder. Then the overlapped ends of the joists are nailed (16d) together. See Fig. 5-22. These nails are driven at an angle, as in Fig. 5-29.

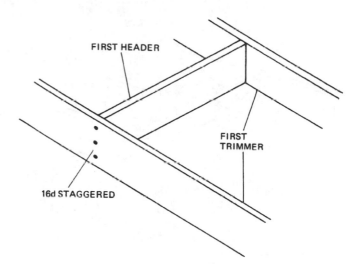

Fig. 5-30 *Nailing the first parts of an opening.*

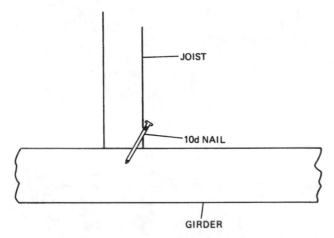

Fig. 5-28 *The joists are toenailed to the girders.*

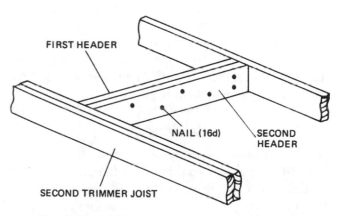

Fig. 5-31 *Add the second header and trimmer joists.*

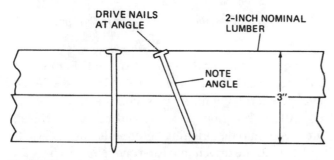

Fig. 5-29 *When nailing joists together, drive nails at an angle. This holds better and the ends do not stick through.*

Nail opening frame A special sequence must be used around the openings. The regular joists next to the opening should not be nailed down. The opening joists are nailed (16d) in place first. These are called *trimmers*.

Then, the first *headers* for openings are nailed (16d) in place. Note that two headers are used. For 2 × 10-inch joists, three nails are used. For 2 × 12-inch lumber, four nails are used. Figure 5-30 shows the spacing of the nails.

Next, short cripple or tail joists are nailed in place. They span from the first header to the joist header. Three 16d nails are driven at each end. Then, the second header is nailed (16d) in place. See Fig. 5-31.

The double trimmer joist is now nailed (16d) in place. These pieces are nailed next to the opening. The nails are alternated top and bottom. See Fig. 5-32. This

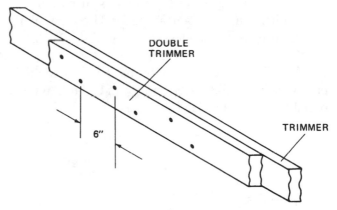

Fig. 5-32 *Stagger nails on double trimmer. Alternate nails on top and bottom.*

finishes the opening. The regular joists next to the opening are nailed in place. Finally, the header is nailed to the joist ends. Three 16d nails are driven into each joist.

Fire-Stops

Fire-stops are short pieces nailed between joists and studs. See Fig. 5-33. They are made of the same boards

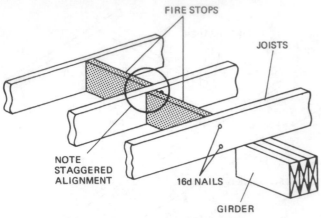

Fig. 5-33 *Fire-stops are nailed in. They keep fire from spreading between walls and floors.*

as the joists. Fire-stops keep fire from spreading between walls and floors. They also help keep joists from twisting and spreading. Fire-stops are usually put at or near the girder. Two 16d nails are driven at each end of the stop. Stagger the boards slightly as shown. This makes it easy to nail them in place.

Bridging

Bridging is used to keep joists from twisting or bending. Bridging is centered between the girder and the header. For most spans, center bridging is adequate. For joist spans longer than 16 feet, more bridging is used. Bridging should be put in every 8 feet. This must be done to comply with most building codes.

Most bridging is cut from boards. It may be cut from either 1-inch or 2-inch lumber. Use the framing square to mark the angles as in Fig. 5-34. With this method, the angle may be found.

A radial arm saw may be used to cut multiple pieces. See Fig. 5-35. Also, a jig can be built to use a portable power saw. See Fig. 5-36.

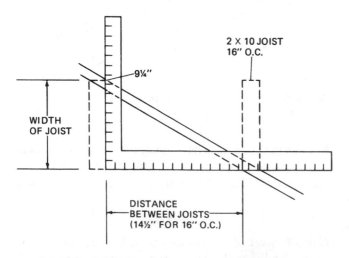

Fig. 5-34 *Use a square to lay out bridging.*

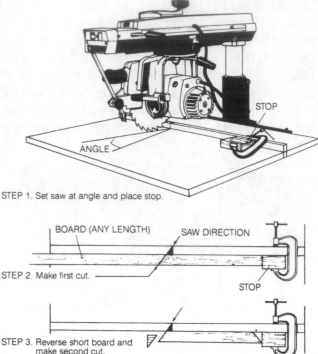

STEP 1. Set saw at angle and place stop.

STEP 2. Make first cut.

STEP 3. Reverse short board and make second cut.

STEP 4. Reverse long board and make cut.

Fig. 5-35 *Radial arm saw set up for bridging.*

Special steel bridging is also used. Figure 5-37 shows an example. Often, only one nail is needed in each end. Steel bridging meets most codes and standards.

All the bridging pieces should be cut first. Nails are driven into the bridging before it is put up. Two 8d or 10d nails are driven into each end. Next, a chalk line is strung across the tops of the joists. This gives a line for the bridging.

The bridging is nailed at the top first. This lets the carpenter space the joists for the flooring when it is laid. The bottoms of the bridging are nailed after the flooring is laid.

The bridging is staggered on either side of the chalk line. This prevents two pieces of bridging from being nailed at the same spot on a joist. To nail them both at the same place would cause the joist to split. See Fig. 5-38.

SUBFLOORS

The last step in making a floor frame is laying the subflooring. Subflooring is also called *underlayment*. The subfloor is the platform that supports the rest of the structure. It is covered with a finish floor material in the living spaces. This may be of wood,

Either way, two kinds of boards are used. Plain boards are laid with a small space between the boards. It allows for expansion. End joints must be made over a joist for support. See Fig. 5-47. Grooved boards are also used. See Fig. 5-48. End joints may be made at any point with grooved boards.

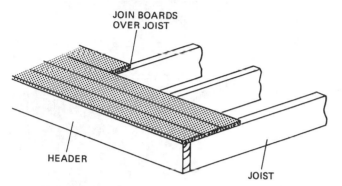

Fig. 5-47 *Plain board subfloor may be laid across joists. Joints must be made over a joist.*

Fig. 5-48 *Grooved flooring is laid across joists. Joints can be made anywhere.*

Nailing Boards are laid from an outside edge toward the center. The first course is laid and nailed with 8d nails. Two nails are used for boards 6 inches wide or less. Three nails are used for boards wider than 6 inches.

The boards are nailed down untrimmed. The ends stick out over the edge of the floor. This is done for both grooved and plain boards. After the floor is done, the ends are sawed off. They are sawed off even with the floor edges.

SPECIAL JOISTS

The carpenter should know how to make special joists. Several types of joists are used in some buildings. Spe-

cial joists are used for overhangs and sunken floors. A sunken floor is any floor lower than the rest. Sunken floors are used for special flooring such as stone. Floors may also be lowered for appearance. Special joists are also used to recess floors into foundations. This is done to make a building look lower. This is called the *low-profile* building.

Overhangs

Overhangs are called *cantilevers*. They are used for special effects. Porches, decks, balconies, and projecting windows are all examples. Figure 5-49A shows an example of projecting windows. Figure 5-49B is a different type of bay. However, both rest on overhanging floor joist systems. Overhangs are also used for "garrison" style houses. When a second floor extends over the wall of the first, it is called a garrison style. See Fig. 5-50A and B.

The longest projection without special anchors is 24 inches. Windows and overhangs seldom extend 24 inches. A balcony, however, would extend more than 24 inches. Thus, a balcony would need special anchors.

Overhangs with joist direction Some overhangs project in the same direction as the floor joists. Little extra framing is needed for this. This is the easiest way to build overhangs. In this method, the joists are simply made longer. Blocking is nailed over the sill with 16d nails. Figure 5-51 shows blocking and headers for this type of overhang. Here the joists rest on the sill. Some overhangs extend over a wall instead of a foundation. Then, the double top plate of the lower wall supports the joists.

Overhangs at angles to joist direction Special construction is needed to frame this type of overhang. It is similar to framing openings in the floor frame. Stringer joists form the base for the subflooring. Stringer joists must be nailed to the main floor joists. See Fig. 5-52. They must be inset twice the distance of the overhang. Two methods of attaching the stringers are used. The first method is to use a wooden ledger. However, this ledger is placed on the top. See Fig. 5-53. The other method uses a metal joist hanger. Special anchors are needed for large overhangs such as rooms or decks.

Sunken Floors

Subfloors are lowered for two main reasons. A finish floor may be made lower than an adjoining finish floor for appearance. Or the subfloor may be made lower to accommodate a finish floor of a different material. The different flooring could be stone, tile, brick, or concrete;

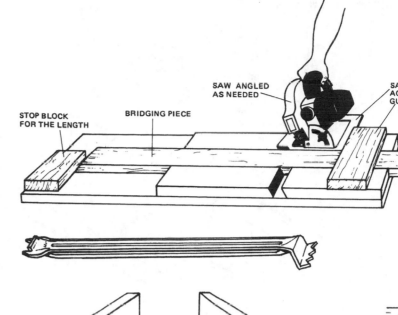

Fig. 5-36 *A jig may be built to cut bridging.*

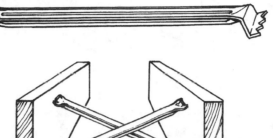

Fig. 5-37 *Most building codes allow steel bridging.*

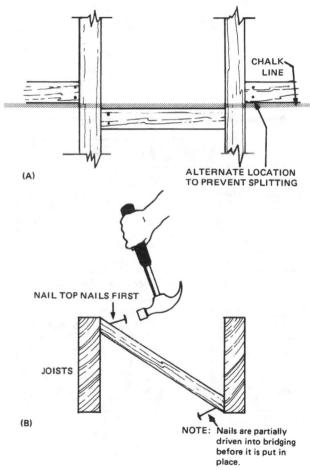

Fig. 5-38 *Bridging pieces are staggered.*

carpeting, tile, or stone. However, the finish floor is added much later.

Several materials are commonly used for subflooring. The most common material is plywood. Plywood should be C–D grade with waterproof or exterior glues. Other materials used are chipboard, fiberboard, and boards.

Plywood Subfloor

Plywood is an ideal subflooring material. It is quickly laid and takes little cutting and trimming. It may be either nailed or glued to the joists. Plywood is very flat and smooth. This makes the finished floor smooth and easy to lay. Builders use thicknesses from ½- to ¾-inch plywood. The most common thicknesses are ½ and ⅝ inch. The FHA minimum is ½-inch-thick plywood.

Plywood as subflooring has fewer squeaks than boards. This is so because fewer nails are required. The squeak in floors is caused when nails work loose. Table 5-1 shows minimum standards for plywood use.

Chipboard and Fiberboard

As a rule, plywood is stronger than other types of underlayment. However, both fiberboard and chipboard

are also used. Chipboard underlayment is used more often. The minimum thickness for chipboard or fiberboard is ⅝ inch. This thickness must also be laid over 16-inch joist spacing. Both chipboard and fiberboard are laid in the same manner as plywood. In any case, the ends of the large sheets are staggered. See Fig. 5-39.

Laying Sheets

The same methods are used for any sheet subflooring. Nails are used most often, but glue is also used. An

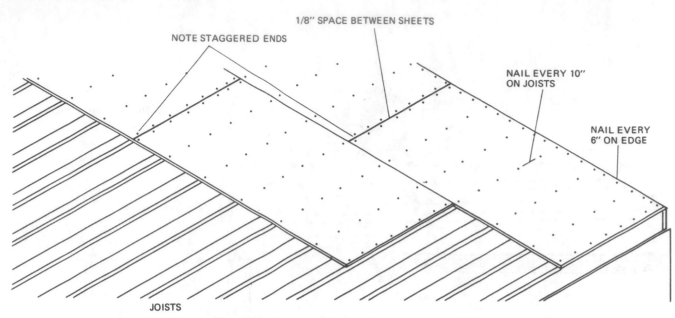

Fig. 5-39 *The ends of subfloor sheets are staggered.*

Labels in figure: NOTE STAGGERED ENDS · 1/8" SPACE BETWEEN SHEETS · NAIL EVERY 10" ON JOISTS · NAIL EVERY 6" ON EDGE · JOISTS

Table 5-1 *Minimum Flooring Standards*

Single-Layer (Resilient) Floor			
Joists, Inches O.C.	Minimum Thickness, Inches	Common Thickness, Inches	Minimum Index
12	19/32	5/8	24/12
16	5/8	5/8 or 3/4	32/16
24	3/4	3/4	48/24
Subflooring with Finish Floor Layer Applied			
Joists, Inches O.C.	Minimum Thickness, Inches	Common Thickness, Inches	Minimum Index
12	1/2	1/2 or 5/8	32/16
16	1/2	5/8	32/16
24	3/4	3/4	48/24

NOTES: 1. C-C grade underlayment plywood.
2. Each piece must be continuous over two spans.
3. Sizes can vary with span and depth of joists in some locations.

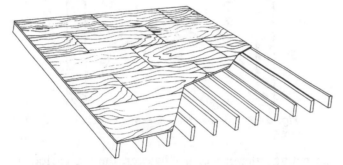

Fig. 5-40 *The long grain runs across the joists.*

outside corner is used as the starting point. The long grain or sheet length is laid across the joists. See Fig. 5-40. The ends of the different courses are staggered. This prevents the ends from all lining up on one joist. If they did, it could weaken the floor. By staggering the end joints, each layer adds strength to the total floor. The carpenter must allow for expansion and contraction. To do this, the sheets are spaced slightly apart. A paper match cover may be used for spacing. Its thickness is about the correct space.

Nailing The outside edges are nailed first with 8d nails. Special "sinker" nails may be used. The outside

nails should be driven about 6 inches apart. Nails are driven into the inner joists about 10 inches apart. See Fig. 5-41. Power nailers can be used to save time, cost, and effort. See Fig. 5-42.

Gluing Gluing is now widely used for subflooring. Modern glues are strong and durable. Glues, also called adhesives, are quickly applied. Glue will not squeak as will nails. Figure 5-43 shows glue being applied to floor joists. Floors are also laid with tongue-

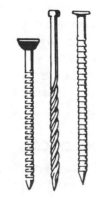

Fig. 5-41 *Flooring nails. Note the "sinker" head on the first nail.*

Fig. 5-42 *Using power nailers saves time and effort. (Duo-Fast)*

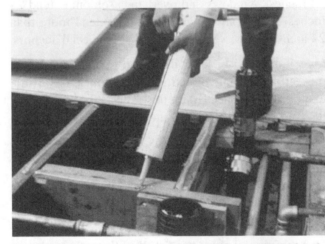

Fig. 5-43 *Subflooring is often glued to the joists. This makes the floor free of squeaks. (American Plywood Association)*

Fig. 5-44 *Plywood subflooring may also have tongue-and-groove joints. This is stronger. (American Plywood Association)*

Fig. 5-45 *A buffer board is used to protect the edges of tongue-and-groove panels. (American Plywood Association)*

and-groove joints (Fig. 5-44). Buffer boards are used to protect the edges of the boards as the panels are put in place. (See Fig. 5-45.)

Board Subflooring

Boards are also used for subflooring. There are two ways of using boards. The older method lays the boards diagonally across the joists. Figure 5-46 shows this. This way takes more time and trimming. It takes a longer time to lay the floor, and more material is wasted by trimming. However, diagonal flooring is still used. It is preferred where wood board finish flooring will be used. This way the finish flooring may be laid at right angles to the joists. Having two layers that run in different directions gives greater strength.

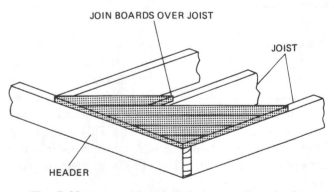

Fig. 5-46 *Diagonal board subfloors are still used today.*

Labels in figure: JOIN BOARDS OVER JOIST · JOIST · HEADER

Today, board subflooring is often laid at right angles to the joists. This is appropriate when the finish floor will be sheets of material.

(A)

(B)

Fig. 5-49 *(A) A bay window rests on overhanging floor joists* (American Plywood Association). *(B) A different type of bay. Both types rest on overhanging floor joist systems.*

(A)

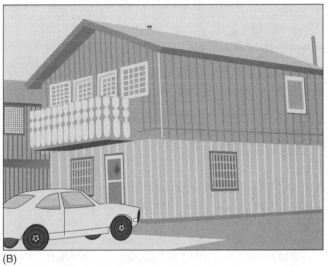

(B)

Fig. 5-50 *(A) Joist protections from the overhang for a garrison-type second story. (B) The finished house.*

These materials are used for appearance or to drain water. However, they are thicker than most finish floors. To make the floor level, special framing is done to lower the subfloor.

The sunken portion is framed like a special opening. First, header joists are nailed (16d) in place. See

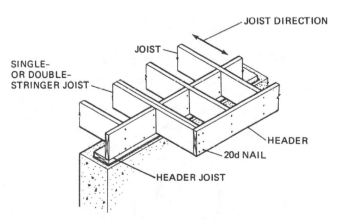

Fig. 5-51 *Some overhangs simply extend the regular joist.* (Forest Products Laboratory)

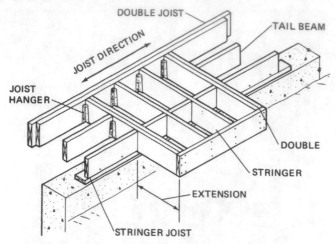

Fig. 5-52 *Frame for an overhang at an angle to the joists.* (Forest Products Laboratory)

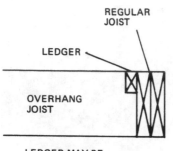

Fig. 5-53 *A top ledger "let in" is a good anchor.*

Fig. 5-54. The headers are not as deep, or wide, as the main joists. This lowers the floor level. To carry the load with thinner boards, more headers are used. The headers are added by spacing them closer together. Double joists are nailed (16d) after the headers.

Low Profiles

The lower-profile home has a regular size frame. However, the subfloor and walls are joined differently. Fig-ure 5-55 shows the arrangement. The sill is below the top of the foundation. The bottom plate for the wall is attached to the foundation. The wall is not nailed to the subfloor. This makes the joists below the common foundation level. The building will appear to be lower than normal.

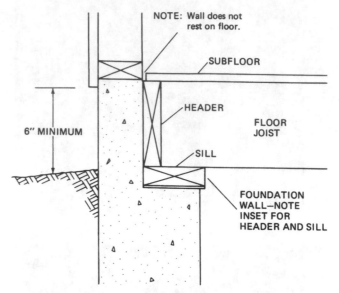

Fig. 5-55 *Floor detail for a low-profile house.*

ENERGY FACTORS

Most energy is not lost through the floor. The most heat is lost through the ceiling. This is so because heat rises. However, energy can be saved by insulating the floor. In the past, most floors were not insulated. Floors over basements need not be insulated. Floors over enclosed basements are the best energy savers.

Floors over crawl spaces should be insulated. The crawl space should also be totally enclosed. The foundation should have ventilation ports. But they should

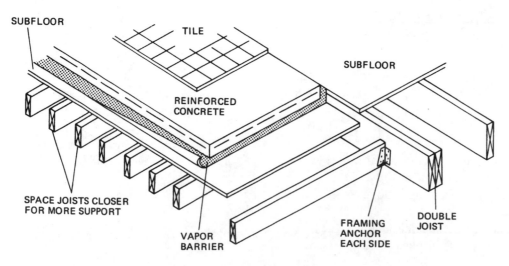

Fig. 5-54 *Details of frame for a sunken floor.*

be closed in winter. The most energy is saved by insulating certain areas. Floors under overhangs and bay windows should be insulated. Floors next to the foundation should also be insulated. The insulation should start at the sill or header. It should extend 12 inches into the floor area. See Fig. 5-56. The outer corners are the most critical areas. But, for the best results, the whole floor can be insulated. Roll or bat insulation is placed between joists and supported. Supports are made of wood strips or wire. Nail (6d) them to the bottom of the joists. See Fig. 5-57.

Fig. 5-57 *Insulation between floor joists should be supported.*

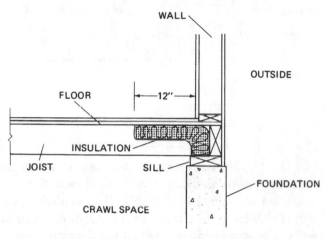

Fig. 5-56 *Insulate the outside floor edges.*

Moisture Barriers

Basements and slabs must have moisture barriers beneath them. Moisture barriers are not needed under a floor over a basement. However, floors over crawl spaces should have moisture barriers. The moisture barrier is laid over the subfloor. See Fig. 5-58. A moisture barrier may be added to older floors below the joists. This may be held in place by either wooden strips or wires. Six-mil plastic or builder's felt is used. See Fig. 5-59.

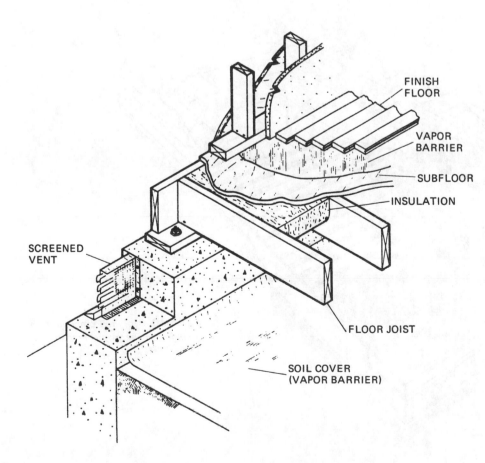

Fig. 5-58 *A moisture barrier is laid over the subfloor above a crawl space.* (Forest Products Laboratory)

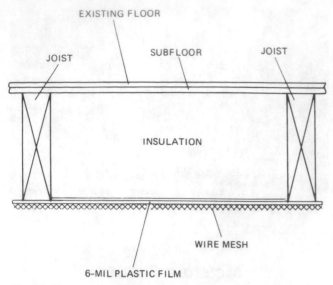

Fig. 5-59 *A moisture barrier may also be added beneath floors.*

Energy Plenums

A *plenum* is a space for controlled air. The air is pressurized a little more than normal. Plenum systems over crawl spaces allow air to circulate beneath floors. This maximizes the heating and cooling effects. Figure 5-60 shows how the air is circulated. Doing this keeps the temperature more even. Even temperatures are more efficient and comfortable.

The plenum must be carefully built. Insulation is used in special areas. See Fig. 5-61. A hatch is needed for plenum floors. The hatch gives access to the plenum area. Access is needed for inspection and servicing. There are no outside doors or vents to the plenum.

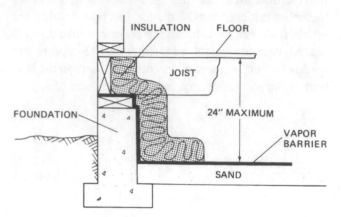

Fig. 5-61 *Section view of the energy plenum.*

The plenum arrangement offers an advantage to the builder: A plenum house can be built more cheaply. There are several reasons for this. The circulation system is simpler. No ducts are built beneath floors or in attics. Common vents are cut in the floors of all the rooms. The system forces air into the sealed plenum

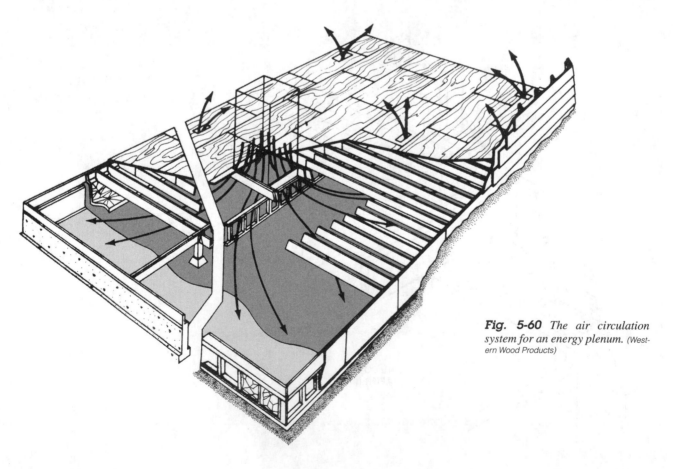

Fig. 5-60 *The air circulation system for an energy plenum.* (Western Wood Products)

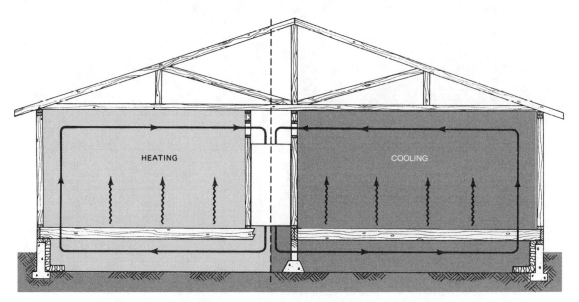

Fig. 5-62 *Conditioned air is circulated through the plenum.* (Western Wood Products)

(see Fig. 5-62). The air does not lose energy in the insulated plenum. The forced air then enters the various rooms from the plenum. The blower unit is in a central portion of the house. The blower can send air evenly from a central area. The enclosed louvered space lets the air return freely to the blower.

Rough plumbing is brought into the crawl space first. Then the foundation is laid. Fuel lines and cleanouts are located outside the crawl space. The minimum clearance in the crawl space should be 18 inches. The maximum should be 24 inches. This size gives the greatest efficiency for air movement.

Foundation walls may be masonry or poured concrete. Special treated plywood foundations are also used. Proper drainage is essential. After the foundation is built, the sill is anchored. Standard sills, seals, and termite shields may be used. The plenum area must be covered with sand. Next, a vapor barrier is laid over the ground and extends up over the sill. See Fig. 5-61. This completely seals the plenum area (Fig. 5-62). Then insulation is laid. Either rigid or batt insulation may be used. It should extend from the sill to about 24 inches inside the plenum. The most energy loss occurs at foundation corners. The insulation covers these corners. Then, the floor joists are nailed to the sill. The joist header is nailed (16d) on and insulated. Subflooring is then nailed to the joists. This completes the plenum. After this, the building is built as a normal platform.

6
CHAPTER

Private Sewage Facilities

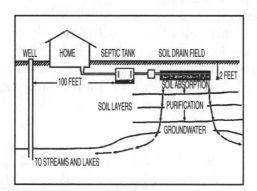

WELL | HOME | SEPTIC TANK | SOIL DRAIN FIELD
100 FEET
SOIL ABSORPTION
2 FEET
SOIL LAYERS | PURIFICATION
GROUNDWATER
TO STREAMS AND LAKES

MANY PEOPLE LIVE IN RURAL AREAS. This choice of living location leads to another expense in the building of a house. Every domicile must have a sewage system as well as a source of water to serve the needs of the occupants. (Chapter 7 covers private water supplies.) Inasmuch as there is no "city sewage system" to serve those outside the limits of most cities, some method of disposal for human wastes must be found. Disposal must be done in a proper manner to prevent the spread of disease and odor.

One of the most readily acceptable means of fulfilling the requirements of modern living and meeting local health department sanitation regulations is the installation of a septic tank. Local (county) regulations will probably govern the minimum size of the septic tank. There is one rule commonly used to figure the storage capacity of the tank: The tank capacity should equal the number of gallons of sewage entering the tank in a 24-hour period. At the rate of 100 gallons per person per day, a septic tank for a four-person household, usually a two-bedroom house, should have a minimum capacity of 400-gallon storage. If a garbage disposer is used, the capacity should be increased by 50 percent. This means a 400- to 600-gallon tank is called for when a four-person household is used. However, in actual practice, the minimum size of any septic tank should be 1000 gallons.

The actual size of the disposal system depends on the number of fixtures served and the permeability of the soil as determined by a percolation test. Sewage disposal systems are designed by sanitary engineers and must be approved and inspected by the health department before being put in use. Consult the building and health codes for specific regulations and requirements.

SEPTIC TANKS AND DISPOSAL FIELDS

The following information is provided for your guidance and planning. An understanding of the septic tank system and its disposal field will aid in living with this type of system.

Septic tanks and disposal fields are used by many kinds of suburban or rural areas that have no service treatment facilities. Septic tanks come in a variety of sizes and shapes. They can be made of precast concrete, fiberglass, or steel. Each type has its advocates, but the precast concrete tank seems to be the longest-lasting, if properly installed.

SEPTIC TANK OPERATION

A septic tank is a covered watertight tank for receiving the discharge from a building sewer and separating out the solid organic matter, which is decomposed and purified by anaerobic bacteria, allowing the clarified liquid to discharge for final disposal. See Fig. 6-1A and B.

The liquid effluent, which is about 70% purified, may flow into one of the following systems:

- A drain field is an open area containing an arrangement of absorption trenches through which effluent from a septic tank may seep or leach into the surrounding soil.

- A seepage pit lined with a perforated masonry or concrete wall is sometimes used as a substitute for a drain field when the soil is absorbent and the highest level of water table is at least 2 feet below the bottom of the pit. See Fig. 6-2.

- A subsurface sand filter consists of distribution pipes surrounded by graded gravel in an intermediate layer

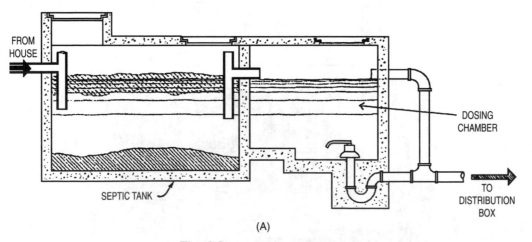

(A)

Fig. 6-1 *(A) Cast concrete tank.*

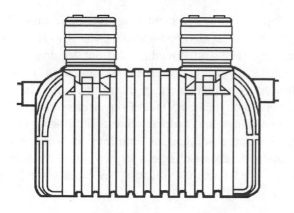

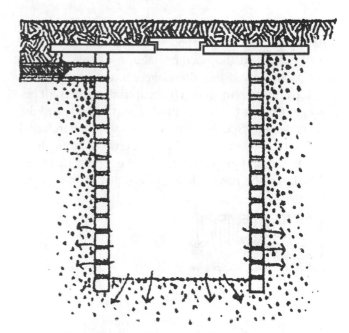

Fig. 6-2 Pit used for sewage disposal purposes.

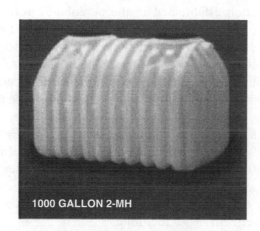

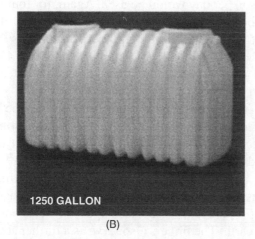

1000 GALLON 2-MH

1250 GALLON

(B)

Fig. 6-1 Continued. (B) Fiberglass tanks.

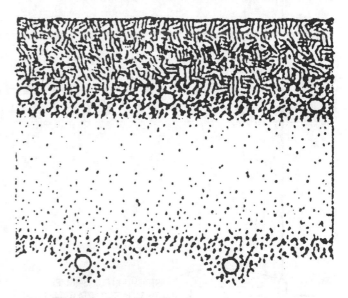

Fig. 6-3 Sand filter. Note the two layers of pipes.

of clean coarse sand, and a system of underdrains to carry off the filtered effluent. Sand filters are used only where other systems are not feasible. See Fig. 6-3.

As mentioned previously, local regulations will probably govern the minimum size of the septic tank. Septic tanks function by a combination of bacterial action and gases. Solids entering the tank drop to the bottom. Bacteria and gases cause decomposition to take place, breaking down the solids into liquids, and in the course of this process the insoluble solids, or sludge, settle to the bottom of the tank. Decomposition in an active tank takes about 24 hours. The sludge builds up on the bottom of the tank so that periodically it must be pumped out. In some cases, it is pumped out once a year. However, in other instances it is only once in 10 years, depending on the usage of the tank. In the cleaning or pumping process, only the sludge on the bottom of the tank should be removed. The crust, formed on the water level at the top of the tank, should not be disturbed. The only exception to this rule occurs when the crust has become coated with grease and the bacterial action of the tank has been destroyed. If this does happen, the crust on top will have to be removed. The bacterial action will begin again when the top is placed on the tank and the tank is sealed. Special compounds can be purchased to hasten the resumption of the bacterial action. However, these compounds are rarely, if ever, needed. Fig. 6-4A shows the rough sketch of a septic tank system.

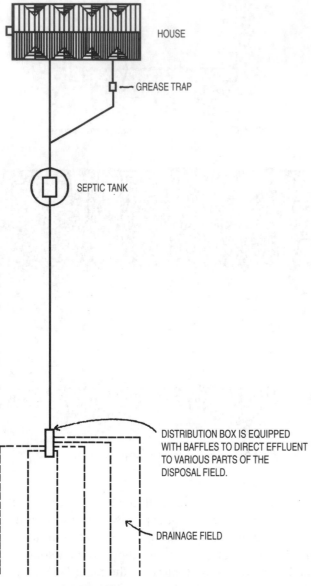

Fig. 6-4A *Sewage disposal system.*

SEPTIC TANK LOCATION

The top of the septic tank should be located at a minimum depth of 12 inches below ground level. The actual depth will probably be somewhat greater due to the depth of the sewer entering the septic tank. The septic tank must also be located at least 100 feet away from any well and downhill, so that all drainage is directed away from the well. Local regulations must be followed as to placement of the septic tank. The system should be located 50 feet from streams and 10 feet from buildings and property lines. The dosing chamber of a large septic tank employs siphonic action to automatically discharge a large volume of effluent when a predetermined quantity has accumulated. See Fig. 6-4B.

SEPTIC TANK DISPOSAL FIELD

The disposal field must be sized using the rate of absorption established in the percolation tests. If the absorption rate is 1 inch in 60 minutes, then 2.35 times the number of gallons of sewage entering the septic tank per day is the number of square feet of trench bottom area needed. At the recommended figure of 100 gallons per day per person, if the absorption rate is 1 inch in 60 minutes and there are four persons in the household, then $2.35 \times 400 =$ square feet of trench bottom needed for the disposal field. If the disposal field has five fingers (trenches), each trench will need 188 square feet of trench bottom; if the trench is 2½ feet (30 inches) wide, then each trench will be 75 feet long.

If the absorption rate established in the tests were 1 inch in 10 minutes, then 0.558×100 gallons per person per day would be used. Again, for four persons in the household, $0.558 \times 400 =$ square feet of trench bottom needed. With five fingers or trenches installed in the disposal field, each trench should have 45 square feet of trench bottom. If the trench is 2½ feet wide, then it should be 18 feet long.

The figures used in these examples may be helpful, but local regulations in your area can vary somewhat. The rules and regulations governing sizes of disposal fields in your area should be followed.

Disposal fields serve two purposes. The fingers of a disposal field provide storage for the discharge of a septic tank until the discharge can be absorbed by the earth. The fingers also serve as a further step in the purification of the discharge through the action of bacteria in the earth. Liquids discharged into the disposal field are disposed of in two ways. One is by evaporation into the air. Sunlight, heat, and capillary action draw the subsurface moisture to the surface in dry weather. See Fig. 6-5A.

In periods of wet or extremely cold weather, the liquids must be absorbed into the earth. The discharge

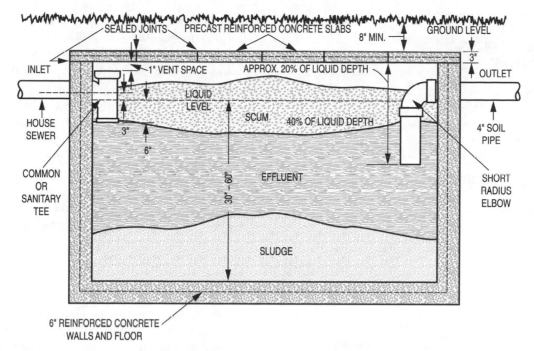

Fig. 6-4B *Sectional view of a typical septic tank.*

from the septic tank should go into a siphon chamber. It is then siphoned into a distribution box. The distribution box directs the discharge so that approximately the same amount of liquid goes into each trench. The drainage field should not run more than 60 feet. Figure 6-5B shows the sewage septic tank disposal field layout.

A siphon, also referred to as a *dosing* siphon, is desirable for this reason: The siphon does not operate until the water level in the siphon chamber reaches a predetermined point. When this level is reached, the siphon action starts and a given number of gallons are discharged into the disposal field. The sudden rush of this liquid into the distribution box and then into the separate fingers of the disposal system ensures that each finger will receive an equal share of the discharge. The discharge thus received by the fingers will have time to be absorbed before another siphon action occurs. A siphon is not essential to the operation of a septic tank and disposal field. However, a siphon will improve the efficiency of the system. The lateral distance between trenches in a disposal field will be governed by ground conditions and local regulations. Construction of the disposal field varies due to soil conditions; basically, a layer of gravel is placed in the bottom of the trench, and then field tile or perforated drain tile is laid on the gravel. If the field tile (farm tile) is used, the tiles should be spaced ¼ inch apart and should slope away from the distribution box at the rate of 4 inches of fall per 100 feet. The space between the

tile should be covered by a strip of heavy asphalt-coated building paper or by a strip of asphalt roofing shingle. See Fig. 6-6.

The trench should then be filled to within 6 inches of ground level with gravel. Topsoil should be added to bring the finished trench to ground level. Since some settling of the topsoil will occur, it is wise to mound the earth slightly over the trenches to allow for settling.

Absorption trenches are 18 to 30 inches wide and 30 inches deep. They contain coarse aggregate and a perforated distribution pipe through which the effluent from a septic tank is allowed to seep into the soil. The distribution pipes should run perpendicular to the slope and at least 2 feet minimum to the water table. See Fig. 6-7.

THE GREASE TRAP

Grease, which is present in dishwashing water, can destroy the action of a septic tank; if it is allowed to get into the disposal field, grease will coat the surface of the earth in the disposal field and prevent absorption. Grease is normal in sewage; the septic tank can dispose of the grease from the kitchen. If grease is trapped before it reaches the septic tank, the life of the septic tank and the disposal field will be greatly prolonged.

A grease trap must be large enough that the incoming hot water will be cooled off as soon as it reaches the grease trap. A 400-gallon septic tank is ideal for

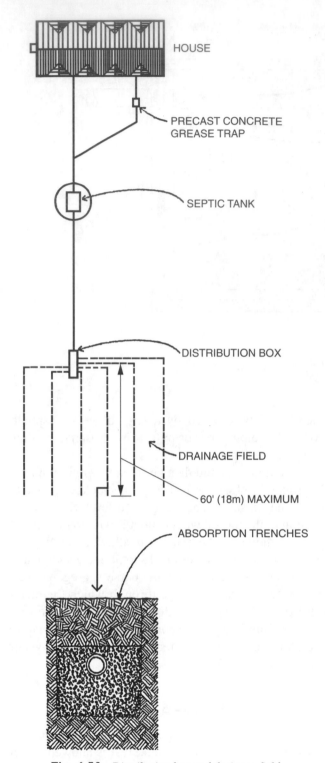

HOUSE

PRECAST CONCRETE
GREASE TRAP

SEPTIC TANK

DISTRIBUTION BOX

DRAINAGE FIELD

60' (18m) MAXIMUM

ABSORPTION TRENCHES

Fig. 6-5A *Distribution box and drainage field.*

use as a grease trap. The water present in this size tank will always be relatively cool, and the grease, present in the dishwashing water, will congeal and rise to the top of the tank. The inlet side and the discharge side of the tank have baffles, or fittings, turned down.

Thus, the water entering and leaving the tank will be trapped. The grease will congeal and float to the top of the tank and relatively clear water will flow out of the grease trap into the septic tank. Drain-cleaning compounds such as sodium hydroxide that contain lye should never be used with a septic tank installation. The drainage piping from the kitchen sink will not be connected to the drainage system of the house when a grease trap is used. The kitchen sink drainage should be piped out separately and the grease trap located as near as possible to the point where the sink drain line exits the house. The top of the grease trap should be within 12 inches of the ground level. A manhole can be extended up to the ground level, with a lightweight locking-type manhole cover; this will provide easy access for skimming off the grease. A periodic check should be made, and the grease skimmed off when it reaches a depth of 3 to 4 inches.

The tile in the disposal field will vary in depth according to local conditions. The septic tank supplier, if unfamiliar with a siphon and siphon chamber, can get this information. If a siphon is not used, the outlet of the septic tank is piped directly to the distribution box. The siphon outlet, if a siphon is used, must be lower than the inlet to the siphon chamber.

NEWER WASTEWATER TREATMENTS

Newer technology provides for a cleaner fluid emission from the wastewater treatment unit. The septic tank has some limitations, and some areas, especially heavily populated zones, no longer allow them to be installed.

One of the newer systems reduces normal household wastewater to clear odorless liquid in just 24 hours. It uses the same process as that in central treatment plants. See Fig. 6-8.

The primary treatment compartment receives the household wastewater and holds it long enough to allow solid matter to settle to the sludge layer at the tank's bottom. Organic solids are broken down here physically and biochemically by anaerobic bacteria— those bacteria that live and work without oxygen. Grit and other untreatable materials are settled out and held back. The partially broken down, finely divided material that is passed on to the aeration compartment is much easier to treat than raw sewage.

In area 2 of Fig. 6-8, the aeration compartment takes the finely divided, pretreated material from the primary compartment and mixes it with activated sludge. It is then aerated. The aerator injects larger quantities of fresh air into the compartment to mix the compartment's entire contents and to provide oxygen for the aerobic digester process.

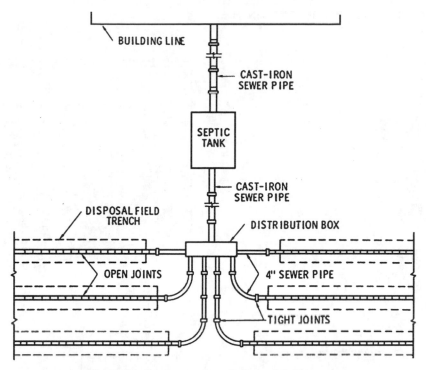

Fig. 6-5B *Sewage septic tank disposal field layout.*

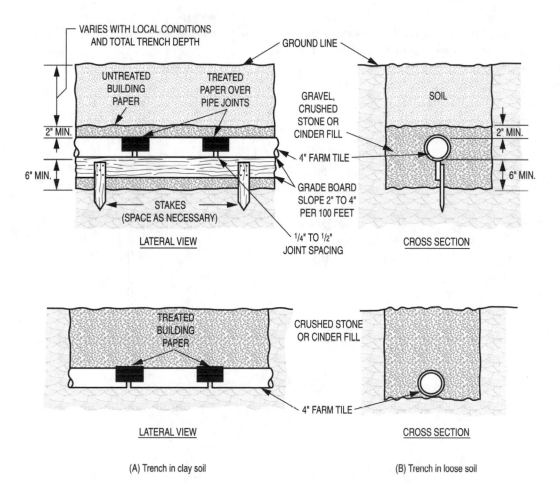

(A) Trench in clay soil

(B) Trench in loose soil

Fig. 6-6 *Trench layouts for various soil conditions.*

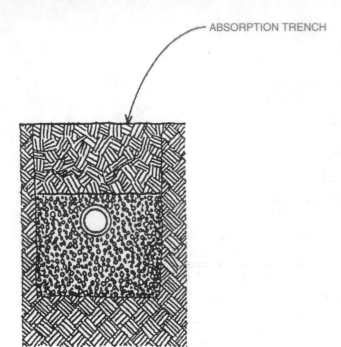

Fig. 6-7 *Absorption trench.*

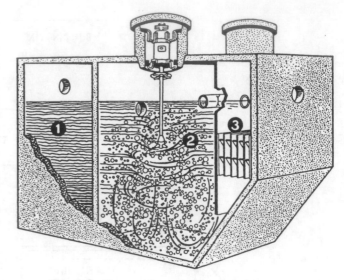

Fig. 6-8 *Home wastewater treatment plant.* (Jet, Inc.)

The aerator (Fig. 6-9) is mounted in a concrete housing that rises to the ground level to give it access to fresh outside air. See Fig. 6-10.

As air is injected into the liquid, the aerator breaks up this air into tiny bubbles so that more air meets the liquid, thus hastening the aerobic digestion process. Aerobic bacteria, which are bacteria that live and work in the presence of oxygen, then use the oxygen in the solution to break down the wastewater completely and convert it to odorless liquids and gases.

The final phase is shown in area 3 of Fig. 6-8. The settling/clarification compartment has a tube settler that eliminates currents and encourages the settling of any remaining particulate material that is returned, by way of the tank's sloping end wall, to the aeration compartment for further treatment. A nonmechanical surface skimmer, operated by hydraulics, skims floating materials from the compartment. The remaining odorless, clarified liquid flows into the final discharge line through the baffled outlet.

This type of home treatment system is often mandated by local health officials where there is a high water table or the soil has poor percolation.

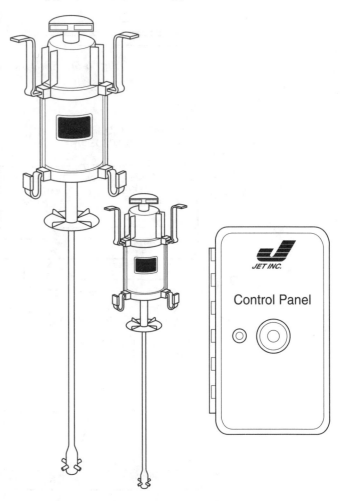

Fig. 6-9 *Aerator and control panel.* (Jet, Inc.)

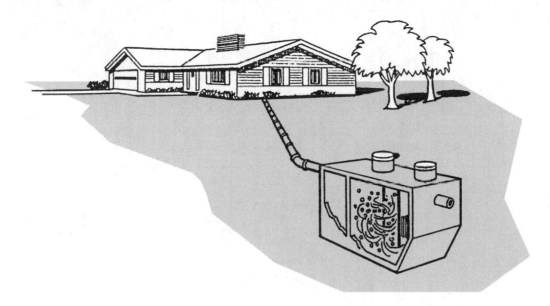

Fig. 6-10 *Home water treatment plant serving a single home.* (Jet, Inc.)

7
CHAPTER

Private Water Systems

WATER SYSTEMS CAN BE PUBLIC OR private. The private water system is the responsibility of the homeowner. It is installed according to the latest rules and regulations furnished by the local health department.

PUBLIC WATER SUPPLIES

The water main is thought of as having always being there. It has been, of course, if you live within a city's limits or in a municipal utility district (MUD) in some states. The water main is the conduit or large pipe through which a public or community water system conveys water to all service connections. Figure 7-1 shows how the typical home installation is made. The water main is usually located in the street—in front of the house—having been located there before the building permits were issued.

Once tapped, the water main furnishes water to the house by way of builder-installed plumbing, a service pipe. This plumbing involves a stop-valve or a place to turn off the water to the house. From there it goes to the curb box, which provides access to a water meter that measures and records the quantity of water that passes through a service pipe. There is usually a control valve for shutting off the water supply to a building in case of an emergency. The service pipe connects a house to a water main. The water main is usually installed by or under the jurisdiction of a public utility.

Once the service has been extended inside the building, a shutoff valve is installed. From there, all the water in the house is controlled by the valve's on/off function.

PRIVATE WATER SYSTEMS

To obtain water for a rural location, out of the reach of city water, a constantly flowing stream must be found or there must be a reliable source of well water.

DRILLING A WELL, BORING A WELL, OR DRIVING A WELL

A well must be drilled and located at least 100 feet away from the house's sewers, septic tanks, and sewage disposal fields; the well should be accessible so as to permit the removal of the well casings or pump for maintenance or repair. Codes are written for well locations and installations and should be checked before you hire someone to produce a well. See Fig. 7-2.

Groundwater is the drinking water source for more than 50 percent of the U.S. population and more than 90 percent of its rural population. In New York State, for example, approximately 3 million rural residents rely on groundwater to supply their drinking water. Most residents with private water supplies also have onsite wastewater treatment systems, commonly called septic systems. Approximately 1.5 million households

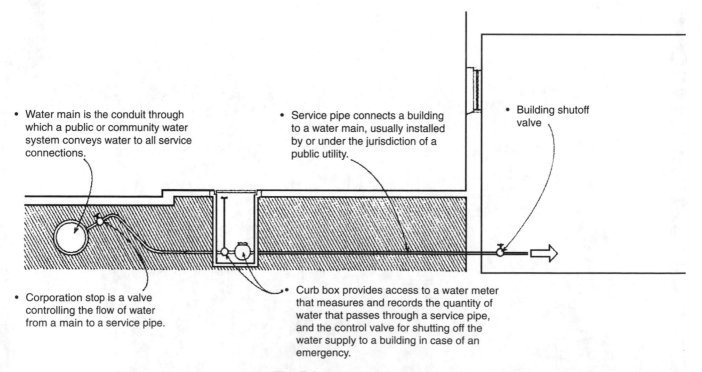

- Water main is the conduit through which a public or community water system conveys water to all service connections.

- Service pipe connects a building to a water main, usually installed by or under the jurisdiction of a public utility.

- Building shutoff valve

- Corporation stop is a valve controlling the flow of water from a main to a service pipe.

- Curb box provides access to a water meter that measures and records the quantity of water that passes through a service pipe, and the control valve for shutting off the water supply to a building in case of an emergency.

Fig. 7-1 *Public water supply.*

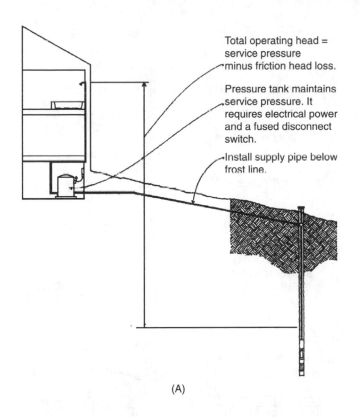

Total operating head = service pressure minus friction head loss.

Pressure tank maintains service pressure. It requires electrical power and a fused disconnect switch.

Install supply pipe below frost line.

(A)

(B)

Fig. 7-2 *(A) Private water system. (B) Drilling a well with modern equipment.*

in New York have some kind of septic system. In most instances they are satisfied with their water supplies— 82 percent are somewhat satisfied or very satisfied.

DRINKING WATER

It is interesting to note that a study of the rural home-owners found that most people are satisfied with their water supplies and do not perceive any problems. However, many fail to

- Take steps to actively protect their water supplies.
- Test their drinking water.
- Pump their septic tanks on a regular basis.

The study showed that upstate New York residents do not employ even low-cost maintenance practices for their water supplies.

Private water supply protection is the responsibility of the individual homeowners. However, only a few local programs sponsor regular testing of private water supplies. This leads to incomplete information and assumptions about the quality of rural drinking water.

Water from wells, springs, lakes, and so forth should be tested for purity. This should be done before the water is used for human consumption. State and/or local boards of health or private laboratories will furnish sterile containers for water samples and test the water for purity. If the water is contaminated, then filters, chlorinators, or iodine feeders can be installed in the system to make the water safe for human consumption.

WELL WATER

Well water, if the source is deep enough, is usually pure, cool, and free of discoloration and taste or odor problems. A sample should be checked for bacteria and chemical content by the local health department before a well is put into operation.

Water must be supplied to a house

- In the correct quantity.
- At the proper flow rate, pressure, and temperature.

For human consumption, water must be potable— free of harmful bacteria—and palatable. To avoid the clogging or corrosion of pipes and equipment, water may have to be treated for hardness or excessive acidity.

Keep in mind that water is consumed by drinking, cooking, cleaning, clothes washing, and bathing. Some heating, ventilation, and air conditioning systems also use water for cooling, heating, and controlling humidity. Fire protection systems store water for extinguishing fires.

The U.S. Environmental Protection Agency (EPA) has a booklet to furnish information to those using private drinking water wells.

Environmental Protection Agency
Ariel Rios Building
1200 Pennsylvania Avenue, NW
Washington, DC 20460
(202) 272-06167

Listed below are the six basic steps you should take to maintain the safety of your drinking water once a system has been installed.

1. Identify potential problem sources.

2. Talk with "local experts."

3. Have your water tested periodically.

4. Have the test results interpreted and explained clearly.

5. Set a regular maintenance schedule for your well, do the scheduled maintenance, and keep accurate, up-to-date records.

6. Remedy any problems.

WATER PRESSURE

Pressure tanks can be used to maintain service pressure. They require electric power to the pump and a fused disconnect switch. For proper operation, the water level in a pressure tank should be based on two-thirds water and one-third air. With the top one-third of the tank filled with air, water is pumped into the tank, and the air is compressed until the top (or high) setting of the pressure switch is reached, usually 40 pounds. The pressure switch then turns off the pump. Water can be used from the tank. As the water is pushed out of the tank, the pressure drops until the low setting of the pressure switch is reached (usually 20 pounds) and the pump starts, beginning the cycle over again. Occasionally, due to the failure of an air volume control or a leak in the pressure tank, the air cushion (the top one-third of the tank) is lost or diminished.

If there is only a very small area of air at the top of the tank, the pressure will drop from 40 pounds to 20 pounds and immediately upon opening of a faucet or valve, the pump will start. If there is no air at the top of the tank, the pump will operate continuously because the pump will be unable to build up enough pressure to cause the pressure switch to turn the pump off. When these conditions occur, the tank is waterlogged. A waterlogged tank should be corrected as soon as possible; the frequent on/off operation of the pressure switch will result in burned contacts on the switch and possible damage to the pump motor. See Fig. 7-3.

PRESSURE TANKS

Pressure tanks with plastic pipe installations require a torque arrester as well as cable guards. The purpose of the pressure tank is to allow an amount of water to be drawn before the pressure drops enough to cause the pump to start. Without a pressure tank, the pump will start and stop continuously when water is drawn. There

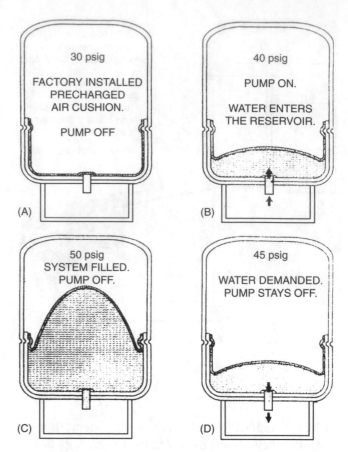

Fig. 7-3 *Sequence of operation for a typical tank.*

are two types of pressure tanks: the standard tank that requires an air volume control and the precharged tank.

• On a standard pneumatic tank system, air is introduced to compensate for that which is absorbed by the water. Each time the pump cycles, air is added to the tank through a bleeder and snifter valve. The excess air is released by a float assembly (air volume control) in the upper side tapping of the tank. See Fig. 7-4.

• In a precharged tank, a flexible diaphragm or bladder separates the air and water areas of the tank. The air chamber is precharged by means of a tire valve with pressure of 2 pounds per square inch less than the cutout pressure of the pump. Because this is not in contact with the water, it cannot be absorbed by the water. Therefore, the original charge of air is never lost. See Fig. 7-5.

In a precharged tank system, none of the fittings for air introduction or air level control are required. See Fig. 7-6. The piping in the well is also different for the two systems. The precharged tank system does not require a bleeder orifice assembly, which simplifies the installation.

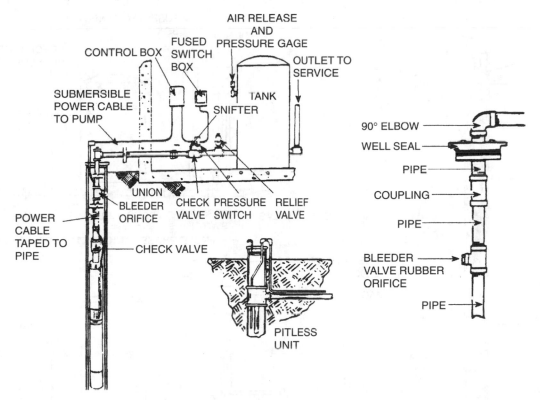

Fig. 7-4 *Typical installation with standard pneumatic tank.*

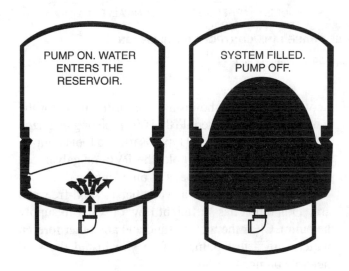

Fig. 7-5 *Bladder operation.*

The tank size should be selected to keep the number of pump starts per day as low as practical for maximum life. Excessive cycling of the motor accelerates motor bearing wear. It also affects spline wear, pump wear, and contact erosion. On single-phase motors, use 100 starts per day (24 hours) as a guide. Motors operating on three-phase current should use 300 starts per day as a rule of thumb.

Pressure tanks that use a permanently sealed-in air charge, similar to the tank shown in Fig. 7-3, can be used instead of the older models. The permanently sealed tank has a number of advantages. Installation is much simpler. There is but one opening in the tank. Waterlogging is eliminated by the sealed-in air cushion, which keeps pump starts to a minimum. This adds to the life of the pump and the control switch. The water reservoir is coated and inside the tank is corrosion-free.

OPERATION OF THE PRESSURE TANK

The four steps in the operation of the pressurized tank are illustrated in Fig. 7-3. In Fig. 7-3A, the tank is precharged with air and the pump is off. In Fig. 7-3B, the pump is on and water enters the reservoir. This compresses the air cushion and raises the air cushion pressure. The tank in Fig. 7-3C shows its filled position and a pressure of 50 psig. This pressure operates a pressure-sensitive switch and turns off the pump. In Fig. 7-3D, the water is being used and the pressure in the air cushion drops. Once it drops to the cut-in point, the pump starts. This is the beginning of a repeating of the cycle. The pressurized tanks are available in sizes from 2 gallons up to 410 gallons. If greater capacity is needed, it is possible to add more tanks. Figure 7-7 shows the many configurations of the single-tank installation.

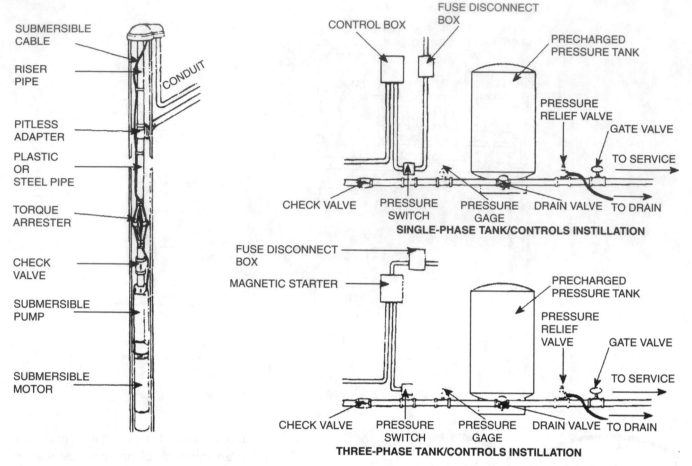

Fig. 7-6 *Typical installation with precharged tank.*

Pressure Switch

The heart of automatic operation is the pressure switch. The pressure switch provides automatic operation by starting when the pressure drops to the switch cut-in setting and stopping when the pressure reaches the switch cutout setting. The pressure switch must be installed as close to the tank as possible. See Figs. 7-6 and 7-4.

Relief Valve (Pressure)

A properly sized pressure relief valve must be installed on any installation where the pump pressure can exceed the pressure tank's maximum working pressure or on systems where the discharge line can be shut off or obstructed. See Fig. 7-8. The relief valve drain port should be piped to a drain.

Keep in mind that not providing a relief valve can cause extreme overpressure, which could result in personal and/or property damage.

PUMP INSTALLATION

After the pump has been tested and the wires have been spliced and wrapped, it is time to drop the pump into the well. Following the installation instructions for the pump chosen, use Schedule 80 PVC pipe or galvanized pipe. If either of these two types is used, a foot clamp or vise will be required to hold the PVC or galvanized pipe when you connect the next pipe length.

Install the pump in a well that is sand-free and straight and that has sufficient flow of water to supply the pump. Clear the well of sand and any other foreign matter with a test pump before you install the submersible pump.

> NOTE: Using the submersible pump to clean the well will void the warranty. When a new well is drilled in an area where sand is a problem, a sand screen must be installed to protect the pump and motor.

Chlorinate the well first. This is done by dropping 24 to 48 HTH chlorine tablets into the well before lowering the pump. This prevents contamination and the growth of iron bacteria. The bacteria could later plug the well and the pump. The chlorinated water is pumped out of the system when testing the pump flow.

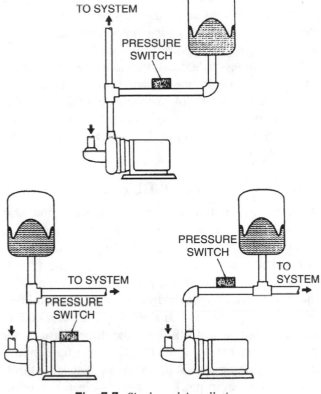

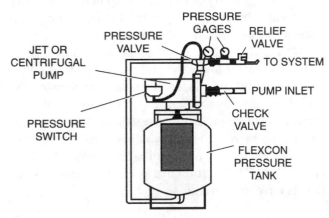

Fig. 7-7 *Single-tank installations.*

Fig. 7-8 *Smart pressure package installations. Note location of the relief valve.* (Flexcon Industries)

Be sure the top edge of the well casing is perfectly smooth, because a sharp or jagged edge can cut or scrape the cable and cause a short.

Install a line check valve within 25 feet of the pump and below the drawdown level of the water supply. The check valve should be the same size as the discharge outlet of the pump or larger.

> NOTE: Use of pipe that is smaller than the discharge tapping of the pump will restrict the capacity of the pump and lower its operating performance.

When connecting the first length of pipe and placing the pump in the well casing, you should take care to maintain a centered pump in the well. It is easier to handle the pump if a short piece of pipe is installed first, rather than a long piece. Install the check valve at the end of the first piece of pipe. Do this prior to lowering the pump into the well. Maintain the alignment as the pump is placed and lowered into the well; a torque arrester is recommended. Position the torque arrester to within 6 inches of the pump discharge and clamp the arrester to the pipe. Wrap the pipe with enough tape at the top and bottom of the torque arrester to keep it from sliding up the pipe while the pump is being lowered into the well.

If not already done, splice the electrical cable to the motor leads. See Fig. 7-9. The cable and ground wire should be taped to the discharge pipe. Tape the cable about 5 feet above the discharge and every 20

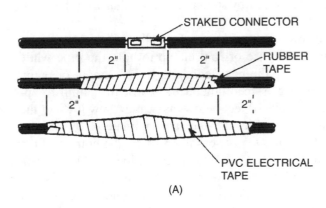

(A)

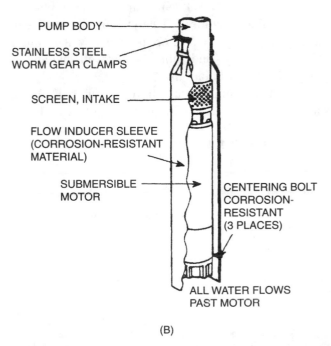

(B)

Fig. 7-9 *(A) Making an electrical splice. (B) Water pump.*

feet thereafter. Install the cable guards if required. This is done to eliminate rubbing against the well casing. Do not let the cable drop over the edge of the well casing. Never allow the weight of the pump to hang on the cable alone.

> WARNING: Since most submersible pump problems are electrical, it is very important that all electrical work be done properly. Therefore, all electrical hookup work or electrical service work should be done only by a qualified electrician or serviceman!

Once the motor (pump) is connected to power, it may be necessary to disconnect the power source before working on or near the motor. Be aware of the shock potential of the control box and the wiring. If the power disconnect is out of sight, be sure to lock it in the open (or off) position and tag it to prevent unexpected application of power by someone not familiar with the repair work going on.

Lower the pump into the well slowly without forcing. Use a vise or foot clamp to hold the pipe while connecting the next length. A boom, tripod, or pump setting rig is recommended. Lower the pump to approximately 10 feet below maximum drawdown of the water if possible, and keep approximately 10 feet from the bottom. *Do not* set the pump on the bottom of the well. Before each new length of pipe is added, attach the coupling to the top of that length of pipe. This will provide a stop for the foot clamp to hold while the next section of pipe is being installed.

On a standard tank with air volume control, a bleeder orifice is required. Install the bleeder orifice in the discharge pipe 5 feet or more below the snifter valve. See Fig. 7-4 and Table 7-1 for well seal and or pitless adapter installations.

All installations should have a well seal. Make sure the seal is seated, and tighten the bolts evenly.

Table 7-1 *Well Seal/Pitless Adapter Installation*

Distance Table	
Tank Size, Gallons	Depth from Horizontal Check Valve to Bleeder Orifice, Feet
42	5
82	10
120	15
220	15
315	20
525	20–35

Note: Installations that use a precharged pressure tank do not require a bleeder orifice.

Be sure to assemble the tee to the pipe above the well seal to prevent dropping the pipe and pump down the well as you lower it. It is important that the well seal and piping be protected from freezing.

On a pitless adapter installation, the connection to the system supply line is made below ground. Install the pitless adapter following the instructions included with the particular brand or design being used in the installation. Follow all applicable state and local plumbing codes.

TEST RUN

After the installation is complete, there is the next step—usually, the one you have been waiting for in this project, the test run. The preliminary test run is done when the pump is at a desired depth and a throttle valve is installed for the preliminary test run. Wire the single-phase motor though the control box, following instructions in the box regarding the color coding of the wires and so forth. Wire the three-phase motor through a magnetic starter. Test the cable for continuity with an ohmmeter. With the pump discharge throttled, run the pump until the water is clear of sand or other impurities. Gradually open the discharge.

Be sure you do not stop the pump before the water runs clear. This may take several hours. If the pump stops with sand in it, it will lock.

If the pump lowers water in the well far enough to lose prime, either lower the pump in the well (if possible) or throttle the discharge to the capacity of the well. If the well is low-capacity, use a low well level control. On three-phase units, establish the correct motor rotation by running in both directions. Change the rotating direction by exchanging any two of the three motor leads. The rotation that gives the greatest water flow is the correct rotation.

PRESSURE TANK INSTALLATION

On a new installation, you have to install the pressure tank along with the pressure switch, pressure gage, pressure relief valve, check valve, gate valves, and unions, as shown in Figs. 7-6 and 7-4.

On replacement pump installations, make sure that the tank system is in good operating condition. A waterlogged tank may cause pump failure.

WATER CONDITIONING EQUIPMENT

Some wells produce water that is usable only with filtering and conditioning. For instance, a well with a high iron content in the water can cause depositions of iron oxide to form in the piping, tanks, water heaters, and water closet tanks. Once these deposits are formed, it is almost impossible to get rid of them. Water pressure tanks, water heaters, and piping may have to be replaced. However, you can install an iron filter to remove the iron before it causes damage in the piping system. Filters can also be installed to remove any objectionable taste and odor from so-called *sulfur* water. Well water, if the source is deep enough, is usually pure and cool, with no discoloration, taste, or odor problems. A sample should be checked for bacteria and chemical content by the local health department before a well is put into permanent operation.

Locating the Equipment

Water conditioning equipment should have its location well planned. It depends on the location and type of house being served. There are a number of ways to connect the conditioner to the filter and to condition the water. Figure 7-10 illustrates a few of the possible locations of the water conditioner.

Select the location of the water softener with care. Various conditions contribute to proper location:

- Locate as close as possible to the water supply source.
- Locate as close as possible to a floor or laundry tube drain.
- Locate the unit in correct relationship to other water conditioning equipment.
- Select a location where the floor is level. If the floor is rough and/or uneven, you can level it by placing the cabinet or tanks on ¾-inch plywood and shim to level as needed.
- Locate the softener in the supply line before the hot water heater. Temperatures above 100°F (38°C) will damage the softener and void the warranty.
- Allow sufficient space around the softener installation for easy servicing.
- Provide a nonswitched 110/120-volt, 60-hertz power source for the control valve.

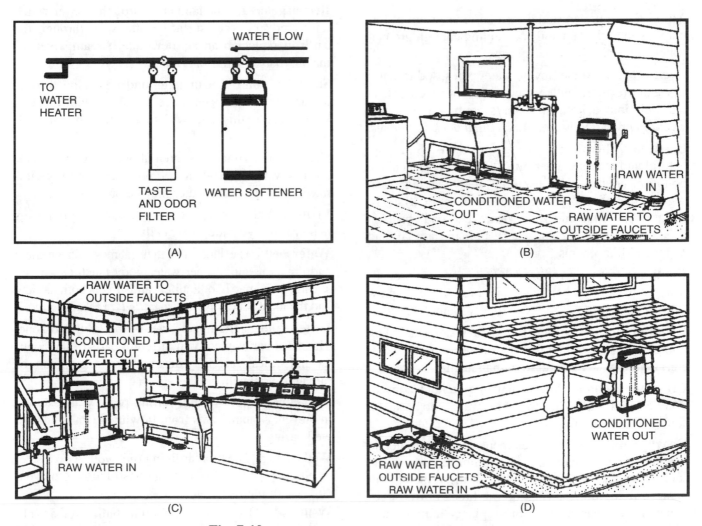

Fig. 7-10 *Typical installations and equipment locations.*

Complete instructions come with the installation whether you buy it at Sears or it is installed by a professional plumber.

EPA regulates public water systems; it does not have the authority to regulate private drinking water wells. Approximately 15 percent of Americans rely on their own private drinking water supplies, and these supplies are not subject to EPA standards, although some state and local governments do set rules to protect users of these wells. Unlike public drinking water systems serving many people, they do not have experts regularly checking the water's source and quality before it is sent to the tap. These households must take special precautions to ensure the protection and maintenance of their drinking water supplies.

DEFINITIONS

The new well owner should have knowledge related to the new water system. Some of the words most often encountered in the process of ownership, repair, and installation are shown here.

Aquifer An underground formation or group of formations in rocks and soils containing enough groundwater to supply wells and springs.

Backflow A reverse flow in water pipes. A difference in water pressures pulls water from sources other than the well into a home's water system, for example, wastewater or floodwater. *Also called back siphonage.*

Bacteria Microscopic living organisms. Some are helpful and others are harmful. "Good" bacteria aid in pollution control by consuming and breaking down organic matter and other pollutants in septic systems, sewage, oil spills, and soils. However, "bad" bacteria in soil, water, or air can cause human, animal, and plant health problems.

Confining layer Layer of rock that keeps the groundwater in the aquifer below it under pressure. This pressure creates springs and helps supply water to wells.

Contaminant Anything found in water (including microorganisms, minerals, chemicals, radionuclides, etc.) that may be harmful to human health.

Cross-connection Any actual or potential connection between a drinking (potable) water supply and a source of contamination.

Heavy metals Metallic elements with high atomic weights, such as mercury, chromium, cadmium, arsenic, and lead. Even at low levels, these metals can damage living things. They do not break down or decompose and tend to build up in plants, animals, and people, causing health concerns.

Leaching field The entire area where many materials (including contaminants) dissolve in rain, snowmelt, or irrigation water and are filtered through the soil.

Microorganisms Very tiny lifeforms such as bacteria, algae, diatoms, parasites, plankton, and fungi. Some can cause disease. *Also called microbes.*

Nitrates Plant nutrients that enter water supply sources from fertilizers, animal feedlots, manures, sewage, septic systems, industrial wastewater, sanitary landfills, and garbage dumps.

Protozoa One-celled animals, usually microscopic, that are larger and more complex than bacteria. *May cause disease.*

Radionuclides Distinct radioactive particles coming from both natural sources and human activities. Can be very long-lasting as soil or water pollutants.

Radon A colorless, odorless, naturally occurring radioactive gas. It is formed by the breakdown or decay of radium or uranium in soil or rocks such as granite. Radon is soluble in water, so well water may contain radon.

Recharge area The land area through or over which rainwater and other surface water soak through the earth to replenish an aquifer, lake, stream, river, or marsh. *Also called a watershed.*

Saturated zone The underground area below the water table where all open spaces are filled with water. A well placed in this zone will be able to pump groundwater.

Unsaturated zone The area above the groundwater level or water table where soil pores are not fully saturated, although some water may be present.

Viruses Submicroscopic disease-causing organisms that grow only inside living cells.

Watershed The land area that catches rain or snow and drains it into a local water body (such as a river, stream, lake, marsh, or aquifer) and affects both its flow and the local water level. *Also called a recharge area.*

Water table The upper level of the saturated zone. This level varies greatly in different parts of the country and varies seasonally depending on the amount of rain and snowmelt.

Well cap A tight-fitting, vermin-proof seal designed to prevent contaminants from flowing down inside the well casing.

Well casing The tubular lining of a well. Also a steel or plastic pipe installed during construction to prevent collapse of the well hole.

Wellhead The top of a structure built over a well. Term also used for the *source of a well or stream.*

8
CHAPTER

Designing and Planning for Solar Heating

SOUTHERN
EXPOSURE

DRAIN

SOLAR SYSTEMS FOR HEATING AND COOLING the house are often thought of as a means of getting something for nothing. After all, the sun is free. All you have to do is devise a system to collect all this energy and channel it where you want it to heat the inside of the house or to cool it in the summer. Sounds simple, doesn't it? It can be done, but it is not inexpensive.

There are two ways to classify solar heating systems. The *passive* type is the simplest. It relies entirely on the movement of a liquid or air by means of the sun's energy. You actually use the sun's energy as a method of heating when you open the curtains on the sunny side of the house during the cold months. In the summer you can use curtains to block the sun's rays and try to keep the room cool. The passive system has no moving parts. (In the example, the moving of the curtains back and forth was an exception.) The design of the house can have a lot to do with this type of heating and cooling. It takes into consideration if the climate requires the house to have an overhang to shade the windows or no overhang so the sun can reach the windows and inside the house during the winter months.

The other type of system is called active. It has moving parts to add to the circulation of the heat by way of pumps to push hot water around the system or fans to blow the heated air and cause it to circulate.

Solar heating has long captivated people who want a care-free system to heat and cool their residences. However, as you examine this chapter, you will find that there is no free source of energy. Some types of solar energy systems are rather expensive to install and maintain. In this chapter you will learn:

• How active and passive systems work

• How cooling and heating are accomplished

• Advantages and disadvantages of an underground house

• How various solar energy systems compare with the conventional methods used for heating and cooling

PASSIVE SOLAR HEATING

Three concepts are used in the passive heating systems: direct, indirect, and isolated gain. See Fig. 8-1. Each of these concepts involves the relationship between the sun, storage mass, and living space.

Indirect Gain

In indirect gain, a storage mass is used to collect and store heat. The storage mass intercedes between the sun and the living space. The three types of indirect

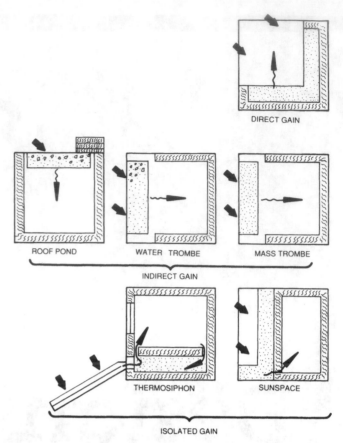

Fig. 8-1 *Three concepts for solar heating: direct gain, indirect gain, and isolated gain.*

gain solar buildings are mass trombe, water trombe, and roof pond. See Fig. 8-1.

Mass trombe buildings involve only a large glass collector area with a storage mass directly behind it. There may be a variety of interpretations of this concept. One of the disadvantages is the large 15-inch-thick concrete mass constructed to absorb the heat during the day. See Fig. 8-2. The concrete has a black coating to aid the absorption of heat. Decorating around this gets to be a challenge. If it is a two-story house, most of the heat will go up the stairwell and remain in the upper bedrooms. Distribution of air by natural convection is viable with this system since the volume of air in the space between the glazing and storage mass is being heated to high temperatures and is constantly trying to move to other areas within the house. The mass can also be made of adobe, stone, or composites of brick, block, and sand.

Cooling is accomplished by allowing the 6-inch space between the mass and the glazed wall to be vented to the outside. Small fans may be necessary to move the hot air. Venting the hot air causes cooler air to be drawn through the house. This will produce some cooling during the summer. The massive wall and

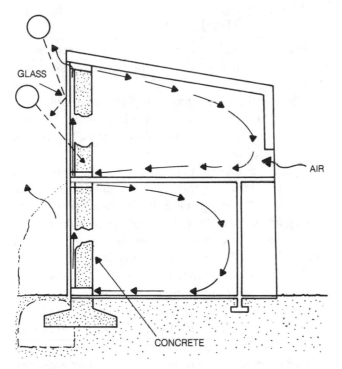

Fig. 8-2 *Indirect gain passive solar heating: mass trombe.*

ground floor slab also maintains cooler daytime temperatures. Trees can be used for shade during the summer and then will drop their leaves during the winter to allow for direct heating of the mass.

A hot-air furnace with ducts built into the wall is used for supplemental heat. Its performance evaluation is roughly 75 percent passive heating contribution. Performance is rated as excellent. For summer, larger vents are needed and in the winter too much heat rises up the open stairwell.

Water trombe buildings are another of the indirect-gain passive heating types. See Fig. 8-3. The buildings have large glazed areas and an adjacent massive heat storage. The storage is in water or another liquid, held in a variety of containers, each with different heat exchange surfaces to storage mass ratios. Larger storage volumes provide greater and longer-term heat storage capacity. Smaller contained volumes provide greater heat exchange surfaces and faster distribution. The trade-off between heat exchange surface versus storage mass has not been fully developed. A number of different types of storage containers, such as tin cans, bottles, tubes, bins, barrels, drums, bags, and complete walls filled with water, have been used in experiments.

A gas-fired hot water heating system is used for backup purposes. That is primarily because this system has a 30 percent passive heating contribution. When fans are used to force air past the wall to improve the heat circulation, it is classified as a *hybrid system*.

The *roof pond* type of building is exactly what its name implies. The roof is flooded with water. See Fig. 8-4. It is protected and controlled by exterior movable insulation. The water is exposed to direct solar gain that causes it to absorb and store heat. Since the heat source is on the roof, it radiates heat from the ceiling to the living space below. Heat is by radiation only. The ceiling height makes a difference to the individual being warmed since radiation density drops off with distance. The storage mass should be uniformly spaced so it covers the entire living area. A hybrid of the passive type must be devised if it is to be more efficient. A movable insulation has to be utilized on sunless winter

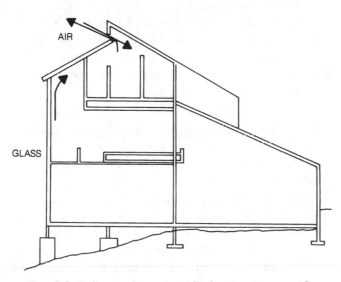

Fig. 8-3 *Indirect gain passive solar heating: water trombe.*

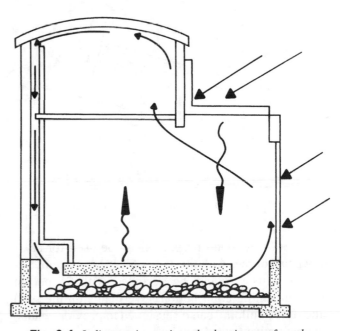

Fig. 8-4 *Indirect gain passive solar heating: roof pond.*

days and nights to prevent unwanted heat losses to the outside. It is also needed for unwanted heat gain in the summer.

This type of system does have some cooling advantages in the summer. It works well for cooling in parts of the country where significant day-to-night temperature swings take place. The water is cooled down on summer evenings by exposure to the night air. The ceiling water mass then draws unwanted heat from the living and working spaces during the day. This takes advantage of the temperature stratification to provide passive cooling.

This type of system has not been fully tested and no specifics are known at this time. It is still being tested in California.

Direct Gain

The direct gain heating method uses the sun directly to heat a room or living space. (See Fig. 8-5.) The area is open to the sun by using a large windowed space so the sun's rays can penetrate the living space. The areas should be exposed to the south, with the solar exposure working on massive walls and floor areas that can hold the heat. The massive walls and thick floors are necessary to hold the heat. That is why most houses cool off at night even though the sun has heated the room during the day. Insulation has to be utilized between the walls and floors and the outside or exterior space. The insulation is needed to prevent the heat loss that occurs at night when the outside cools down.

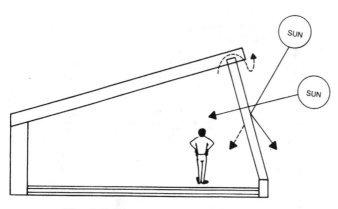

Fig. 8-5 *Direct gain passive solar heating.*

Woodstoves and fireplaces can be used for auxiliary heat sources with this type of heating system. Systems of this type can be designed with 95 percent passive heating contribution. Overhangs on the south side of the building can be designed to provide shading against unwanted solar gain.

Isolated Gain

In this type of solar heating the solar collection and storage are thermally isolated from the living spaces of the building.

The *sunspace* isolated gain passive building type collects solar radiation in a secondary space. The isolated space is separate from the living space. This design stores heat for later distribution.

This type of design has some advantages over the others. It offers separation of the collector-storage system from the living space. It is midway between the direct gain system where the living space is the collector of heat, and a mass or water trombe system that collects heat indirectly for the living space.

Part of the design may be an atrium, a sun porch, a greenhouse, and a sunroom. The southern exposure of the house is usually the location of the collector arrangement. A dark tile floor can be used to absorb some of the sun's heat. The northern exposure is protected by a berm and a minimum of window area. The concrete slab is 8 inches thick and will store the heat for use during the night. During the summer the atrium is shaded by deciduous trees. A fireplace and small central gas heater are used for auxiliary heat. Performance for the test run has reached the 75 to 90 percent level.

The *thermosiphon* isolated gain passive building type generates another type of solar heated building. See Fig. 8-6. In this type of solar heating system the collector space is between the direct sunshine and the living space. It is not part of the building. A thermosiphoning heat flow occurs when the cool air or liquid naturally falls to the lowest point—in this case, the collectors. Once heated by the sun, the heated air or liquid rises up into an appropriately placed living space, or it can be moved to a storage mass. This causes a somewhat cooler air or liquid to fall again. This movement causes a continuous circulation to begin. Since the collector space is completely separate from the building, the thermosiphon system resembles an active system. However, the advantage is that there are no external fans or blowers needed to move the heat transfer medium. The thermosiphon principle has been applied in numerous solar domestic hot water systems. It offers good potential for space heating applications.

Electric heaters and a fireplace with a heatilator serve as auxiliary sources. The south porch can be designed to shade the southern face with overhangs to protect the *clerestory* windows. Clerestory windows are those above the normal roofline such as in Fig. 8-6. Cross-ventilation is provided by these windows.

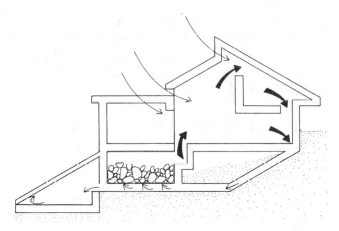

Fig. 8-6 *Isolated gain passive heating system: thermosiphon.*

Time Lag Heating

Time lag heating was used by the American Indians of the Southwest where it got extremely hot in the day and cooled down during the night. The diurnal (day-to-night) temperature conditions provided a clear opportunity for free or natural heating by delaying and holding daytime heat gain for use in the cool evening hours.

In those parts of the country where there are significant (20 to 35 degrees F) day-to-night temperature swings, a building with thermal mass can allow the home itself to delay and store external daytime heating in its walls. The captured heat is then radiated to the building interior during the cool night. Internal heat gains come from people, lights, and appliances. This heat can also be absorbed and stored in the building structure.

Massive or heavy construction of walls, floors, and ceilings are used for this type of solar heating. Because it is dense, concrete, stone, or adobe has the capacity to hold heat. As the outside temperature rises during the day, so does the temperature of the building surface. The entire wall section heats up and will gradually release the stored heat to the room by radiation and convection. Two controls can make time lag heating systems most effective for passive heating. Two, four, eight, even twelve hours of delay can be guaranteed by building walls of the right thickness and density. Choosing the right material and thickness can allow you to control what hour in the evening you begin heating. Exterior or sheathing insulation can prevent the heat storage wall from losing its carefully gained heat to the outside, offering more passive heat to the inside. For additional heat, winter sunshine can also be collected and stored in massive walls, provided adequate shading is given to these walls to prevent overheating in the summer.

Uninsulated massive walls can cause problems with the auxiliary heating system if improperly used. In climates where there is no day-to-night temperature swing, uninsulated massive walls can cause problems. Continually cold or continually hot temperatures outside will build up in heavy exterior walls and will draw heat from the house for hours until these walls have been completely heated from the inside.

This type of house has been built and tested in Denver, Colorado with a 65 percent passive heating contribution and 60 percent passive cooling contribution.

Underground Heating

The average temperature underground, below the frostline, remains stable at approximately 56 degrees. This can be used to provide effective natural heating as outside temperatures drop below freezing. The massiveness of the earth itself takes a long time to heat up and cool down. Its average annual temperature ranges between 55 and 65 degrees, with only slight increases at the end of summer and slight decreases at the end of winter. In climates with severe winter conditions or severe summer temperatures, underground construction provides considerably improved outside design temperatures. It also reduces wind exposure. The underground building method removes most of the heating load for maximum energy conservation.

One of the greatest disadvantages of the underground home is the humidity. If you live in a very humid region of the country, excessive humidity, moisture, and mildew can present problems. This is usually no problem during the winter season, but can become serious in the summer. Underground buildings cannot take maximum advantage of comfortable outside temperatures. Instead, they are continuously exposed to 56 degree ground temperatures. So, if you live in a comfortable climate, there is no need to build underground. You could, however, provide spring and fall living spaces outside the underground dwelling and use it only for summer cooling and winter heat conservation. Make sure you do not build on clay that swells and slides. And do not dig deep into slopes without shoring against erosion.

An example is a test building that was constructed in Minneapolis. The building was designed to be energy efficient. It was also designed to use the passive cooling effects of the earth. Net energy savings over a conventional building are expected to be 80 to 100 percent during the heating period and approximately 45 percent during the cooling period.

PASSIVE COOLING SYSTEMS

There are six passive cooling systems. These include natural and induced ventilation systems, desiccant systems, and evaporative cooling systems, as well as the passive cooling that can be provided by night sky temperature conditions, diurnal (day-to-night) temperature conditions, and underground temperature conditions.

Natural Ventilation

Natural ventilation is used in climates where there are significant summer winds and sufficient humidity (more than 20 percent) so that the air movement will not cause dehydration.

Induced Ventilation

Induced ventilation is used in climate regions that are sunny but experience little summer wind activity.

Desiccant Cooling

Another name for desiccant cooling is *dehumidification*. This type of cooling is used in climates where high humidities are the major cause of discomfort. Humidities greater than 70 or 80 percent RH (relative humidity) will prevent evaporative cooling. Thus, methods of drying out the air can provide effective summer cooling.

This type of cooling is accomplished by using two desiccant salt plates for absorbing water vapor and solar energy for drying out the salts. The two desiccant salt plates are placed alternately in the living space, where they absorb water vapor from the air, and in the sun, to evaporate this water vapor and return to solid form. The salt plates may be dried either on the roof or at the southern wall, or alternately at east and west walls responding to morning and afternoon sun positions. Mechanized wheels transporting wet salt plates to the outside and dry salt plates to the inside could also be used.

Evaporative Cooling

This type of cooling is used where there are low humidities. The addition of moisture by using pools, fountains, and plants will begin an evaporation process that increases the humidity but lowers the temperature of the air for cooling relief.

Spraying the roof with water can also cause a reduction in the ceiling temperature and cause air movement to the cooler surface. Keep in mind that evaporative cooling is effective only in drier climates. Water has to be available for make up of the evaporated moisture. The atrium and the mechanical coolers should be kept out of the sun since it is the air's heat you are trying to use for the evaporation process, not solar heat. For the total system design, the pools of water, vegetation, and fountain court should be combined with the prevailing summer winds for efficient distribution of cool, humidified air.

Night Sky Radiation Cooling

Night sky radiation is dependent on clear nights in the summer. It involves the cooling of a massive body of water or masonry by exposure to a cool night sky. This type of cooling is most effective when there is a large day-to-night temperature swing. A clear night sky in any climate will act as a large heat sink to draw away the daytime heat that has accumulated in the building mass. A well-sized and exposed body of water or masonry, once cooled by radiation to the night sky, can be designed to act as a cold storage, draining heat away from the living space through the summer day and providing natural summer cooling.

The roof pond is one natural conditioning system that offers the potential for both passive heating and passive cooling. The requirements for this system involve the use of a contained body of water or masonry on the roof. This should be protected when necessary by moving insulation or by moving the water. The house has conventional ceiling heights for effective radiated heating and cooling. During the summer when it is too hot for comfort, the insulating panels are rolled away at night. This exposes the water mass to the clear night sky, which absorbs all the daytime heat from the water mass and leaves the chilled water behind. During the day the insulated panels are closed to protect the roof mass from the heat. The chilled storage mass below absorbs heat from the living spaces to provide natural cooling for most or all of the day.

This type of cooling offers up to 100 percent passive, nonmechanical air conditioning.

Time Lag Cooling

This type of cooling has already been described in the section on time lag heating. It is used primarily in the climates that have a large day-to-night temperature swing. The well-insulated walls and floor will maintain the night temperature well into the day, transmitting little of the outside heat into the house. If 20-inch eaves over all the windows are used, they can exclude most of the summer radiation, thereby controlling the direct heat gain.

Underground Cooling

Underground cooling takes advantage of the fairly stable 56-degree temperature conditions of the earth below the frostline. The only control needed is the addition of perimeter insulation to keep the house temperatures above 56 degrees F. In climates of severe summer temperatures and moderate-to-low humidity levels, underground construction provides stable and cool outside design temperatures as well as reduced sun exposure to remove most of the cooling load for maximum energy conservation.

ACTIVE SOLAR HEATING SYSTEMS

Active solar systems are modified systems that use fans, blowers, and pumps to control the heating process and the distribution of the heat once it is collected.

Active systems currently are using the following six units to collect, control, and distribute solar heat.

Unit	Function
1. Solar collector	Intercepts solar radiation and converts it to heat for transfer to a thermal storage unit or to the heating load.
2. Thermal storage unit	Can be either an air or liquid unit. If more heat than needed is collected, it is stored in this unit for later use. Can be either liquid, rock, or a phase change unit.
3. Auxiliary heat source	Used as a backup unit when there is not enough solar heat to do the job.
4. Heat distribution system	Depending on the systems selected, these could be the same as those used for cooling or auxiliary heating.
5. Cooling distribution system	Usually a blower and duct distribution capable of using air or liquid directly from either the solar collector or the thermal storage unit.

Operation of Solar Heating Systems

It would take a book in itself to examine all the possibilities and maybe a couple of volumes more to present details of what has been done to date. Therefore, it is best to take a look at a system that is commercially available from a reputable firm that has been making heating and cooling systems for years. See Fig. 8-7.

Domestic Water Heating System

The domestic water heating system uses water heated with solar energy. It is more economically viable than whole-house space heating because hot water is required all year-round. The opportunity to obtain a return on the initial investment in the system every day of the year is a distinct economic advantage. Only moderate collector temperatures are required

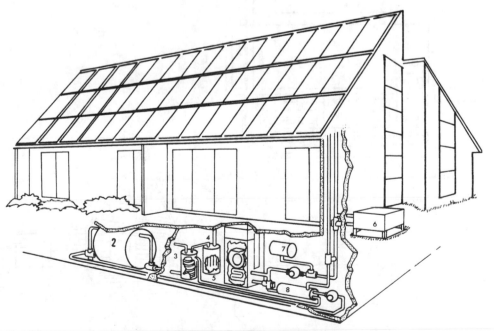

Fig. 8-7 *Components of a liquid to air solar system.* (Lennox Furnace Co.)

Key components in a liquid air solar system: 1. Solar collectors, 2. Storage tank, 3. Hot water heat exchanger, 4. Hot water holding tank, 5. Space heating coil, 6. Purge coil (releases excess solar heat), 7. Expansion tank, 8. Heat exchanger.

to cause the system to function effectively. Thus domestic water can be heated during less than ideal weather conditions.

Indirect Heating/Circulating Systems

Indirect heating systems circulate antifreeze solution or special heat transfer fluid through the collectors. This is primarily to overcome the problem of draining liquid collectors during periods of subfreezing weather. See Fig. 8-8. Air collectors can also be used. As a result, there is no danger of freezing and no need to drain the system.

Circulating a solution of ethylene glycol and water through the collector and a heat exchanger is one means of eliminating the problem of freezing. See Fig. 8-9. Note that this system requires a heat exchanger and an additional pump. The heat exchanger permits the heat in the liquid circulating through the collector to be transferred to the water in the storage tank. The extra pump is needed to circulate water from the storage tank through the heat exchanger.

The extra pump can be eliminated if:

- The heat exchanger is located below the storage tank.
- The pipe sizes and heat exchanger design permit thermosiphon action to circulate the water.
- A heat exchanger is used that actually wraps around and contacts the storage tank and transfers heat directly through the tank wall.

Safety is another consideration in the operation of this type of system. Two major problems might develop with liquid solar water heaters:

1. Excessive water may enter the domestic water service line.
2. High temperature–high pressure may damage collectors and the storage unit. See Fig. 8-10.

If you want to prevent the first problem, you can add a mixing valve between the solar storage tank and the conventional water heater. See Fig. 8-11.

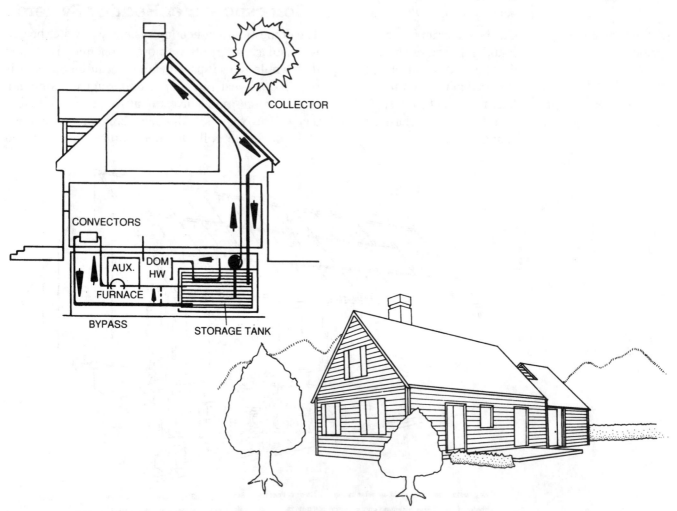

Fig. 8-8 *Collectors on the roof heat water that is circulated throughout the house.*

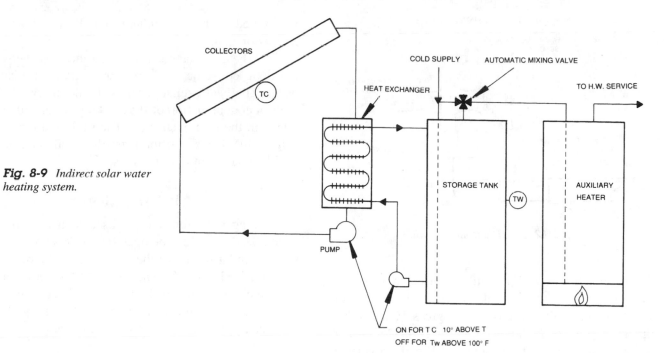

Fig. 8-9 *Indirect solar water heating system.*

COLLECTORS
TC
HEAT EXCHANGER
COLD SUPPLY
AUTOMATIC MIXING VALVE
TO H.W. SERVICE
STORAGE TANK
TW
AUXILIARY HEATER
PUMP
ON FOR T C 10° ABOVE T
OFF FOR Tw ABOVE 100° F

GASKET SEAL
COVER MOUNTING SCREW
ALUMINUM COVER FRAME
FLOW TUBE MANIFOLD
PIPE CONNECTION
MOUNTING BRACKETS
ABSORBER PLATE
INSULATION
RUBBER PADS
GLASS COVERS
COPPER FLOW TUBES
CABINET

(A)

TEMPERED GLASS
ABSORBER PLATE
AIR CHANNEL
HOT AIR
140° F
INSULATION
MANIFOLD AREA
COLD AIR   70° F

(B)

Fig. 8-10 *(A) Liquid collector. (B) Air collector.*

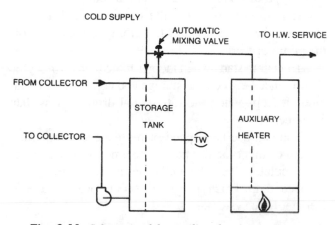

COLD SUPPLY
AUTOMATIC MIXING VALVE
TO H.W. SERVICE
FROM COLLECTOR
STORAGE TANK
AUXILIARY HEATER
TW
TO COLLECTOR

Fig. 8-11 *Schematic of the auxiliary heating equipment.*

Cold water is blended with hot water in the proper proportion to avoid excessive supply temperature. The mixing valve is sometimes referred to as a tempering valve. Figure 8-12 shows the details of a typical connection for a tempering valve.

You can avoid excessive pressure in the collector loop carrying the antifreeze or heat transfer solution by installing a pressure-relief valve in the loop. Set the valve to discharge at anything above 50 psi. The temperature of the liquid may hit 200 degrees F, so make sure the relief valve is connected to an open drain. The fluid is unsafe and contaminated, so keep that in mind when disposing of it.

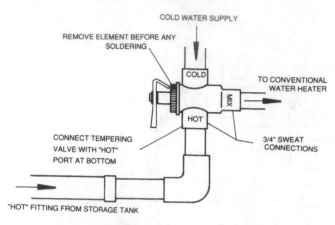

Fig. 8-12 *A typical tempering valve.*

A temperature and pressure-relief valve is usually installed on the storage tank to protect it. Whenever water in the tank exceeds 210 degrees F, the valve opens and purges the hot water in the tank. Cold water automatically enters the storage tank and provides a heating load for the collector loop. This cools down the system. Figure 8-13 gives examples of both safety devices installed in the system.

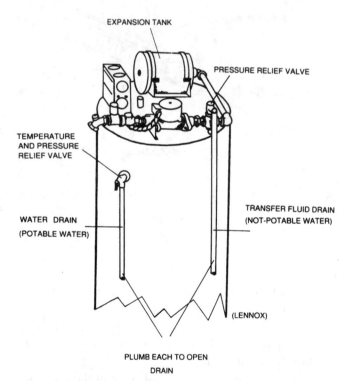

Fig. 8-13 *Note the locations of the safety valves on the heater tank and the expansion tank.*

The collector loop expansion tank (Fig. 8-13) is required to absorb the expansion and contraction of the circulating fluid as it is heated and cooled. Any loop not vented to the atmosphere must be fitted with an expansion tank.

The heat exchanger acts as an interface between the toxic collector fluid and the *potable* (drinkable) water. The heat exchanger must be double-walled to prevent contamination of the drinking water if there is a leak in the heat exchanger. The shell and tube type (Fig. 8-14) does not often meet the local code or the health department requirements.

Air Transfer

Air-heating collectors can be used to heat domestic water. See Fig. 8-15. The operation of this type of system is similar to that of the indirect liquid circulation system. The basic difference is that a blower or fan is used to circulate the air through the collector and heat exchanger rather than a pump to circulate a liquid.

The air transfer method has advantages:

- It does not have any damage due to a liquid leakage in the collector loop.
- It does not have to be concerned with boiling fluid or freezing in the winter.
- It does not run the risk of losing the expensive fluid in the system.

It does have some disadvantages over the liquid type of system.

- It requires larger piping between the collector and heat exchanger.
- It requires more energy to operate the circulating fan than it does for the water pump.
- It needs a slightly larger collector.

Cycle Operation

The indirect and direct water heating systems need some type of control. A differential temperature controller is used to measure the temperature difference between the collector and the storage. This controls the pump operation.

The pump starts when there is more than a 10-degree F difference between the storage and collector temperatures. It stops when the differential drops to less than 3 degrees F.

You can use two-speed or multispeed pumps in a system of this type to change the amount of water being circulated. As solar radiation increases, the pump is speeded up. This type of unit also improves the efficiency of the system.

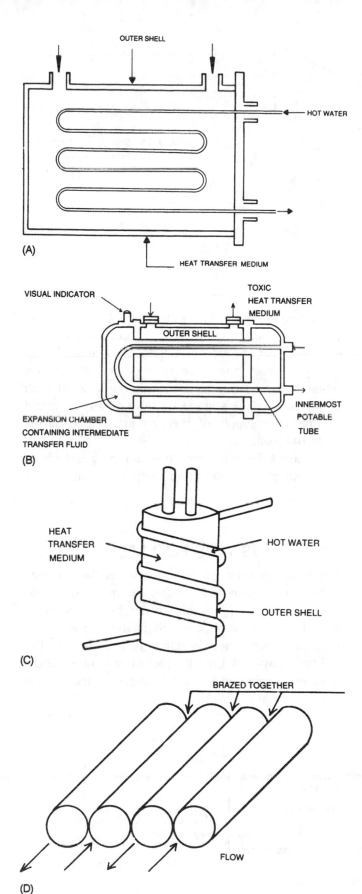

(A)

(B)

(C)

(D)

Fig. 8-14 *Heat exchanger designs.*

Designing the Domestic Water Heating System

Any heating job requires that you know the number of Btus (British Thermal Units) needed to heat a space. The type of heating system we are designing is no exception.

Table 8-1 shows minimum property standards for solar systems as designated by the United States Department of Housing and Urban Development. Note that the minimum daily hot water requirements for various residence and apartment occupancy are listed. For example, a two-bedroom home with three occupants should be provided with equipment that can provide 55 gallons per day of hot water. Many designers simply assume 20 gallons per day per person, which results in slightly higher requirements than those listed in Table 8-1.

Another important consideration in sizing the solar domestic hot water system is the required change in the temperature of incoming water. The water supplied by a public water system usually varies from 40 to 75 degrees F, depending on location and season of the year. A telephone call to your local water utility will provide the water supply temperature in your area. Generally, the desired supply hot water temperature is from 140 to 160 degrees F. Knowing these two temperatures and the volume of water required enables you to calculate the Btu requirement for domestic hot water. Figure 8-16 shows how to calculate the Btu requirements for heating domestic hot water.

To find the required collector area needed to provide some portion of the Btu load, you can use a number of methods. Figure 8-17 shows the location of an add-on collector. One rule of thumb is: the amount of solar energy available at midlatitude in the continental United States is approximately equal to 2000 Btus per square foot per day. Assuming a collector efficiency of 40 percent, 800 Btus per square foot per day can be collected with a properly installed collector. Using the example in Fig. 8-17, the collector should contain approximately 137 square feet. That is found by 109,956 Btus per day divided by 800, which equals 137. The higher summer radiation levels and warmer temperatures would cause an excess capacity most of the time. A more practical approach is to provide nearly 100% solar hot water in July, which might then average out to 70% contribution for the year. Thus a collector area of 0.7 by 137, or about 96 square feet, might be a more realistic installation. See Fig. 8-18.

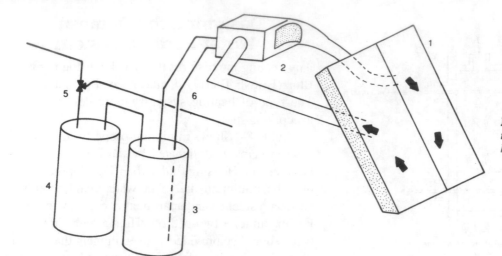

Fig. 8-15 *Schematic of air transfer medium solar water heating system.*

Table 8-1 *Daily Hot Water Usage (140°F) for Solar System Design*

Category	One- and Two-Family Units and Apartments up to 20 Units				
Number of people	2	3	4	5	6
Number of bedrooms	1	2	3	4	5
Hot water per unit (gallons per day)	40	55	70	85	100

The rule of thumb sizing procedures used here assumes that the collector is installed facing due south and inclined at an angle equal to the local latitude plus 10 degrees. Modification of these optimum collector installation procedures will reduce the effectiveness of the collector. In case the ideal installation cannot be achieved, it will be necessary to increase the size of the collector to compensate for the loss in effectiveness.

Other Components

The components for solar domestic hot water heating systems are available in kits prepackaged with instruc-

tions. See Fig. 8-19. These eliminate the need to size the storage tank, expansion tank, and pump. If you wish to select individual components, it will be necessary to make the same type of calculations for whole-house heating to determine the sizes of such components. Tank storage would typically be based on one day's supply of energy that is based on the daily Btu load.

Figure 8-20 illustrates a typical piping and wiring arrangement for a solar water heating system.

IS THIS FOR ME?

The basic question for everyone is: Is this for me? What are the economics of the system? Most manufacturers of packaged solar water heating systems provide some type of economic analysis to assist the installation contractors in selling their customers. For example, pay-back time for fuel savings to equal the total investment may be as little as six to nine years at the present time.

Example Problem:

Given that a family of six people live in a home where the incoming water temperature is 40°F and the requirement is for 150°F hot water, calculate the BTU requirement.

1. Compute: Water requirements = Number of people × 20 gallons per day
   = 6 × 20
   = 120 gallons of hot water used per day

Because a gallon of water weighs 8.33 pounds, the BTU requirement per day for hot water can be found by:

2. Compute: Heat Required = gallons per day × 8.33 × temperature rise
   = (120 × 8.33) (150 − 40)
   = 109,956 BTU per day

Fig. 8-16 *Circulating Btu requirements.*

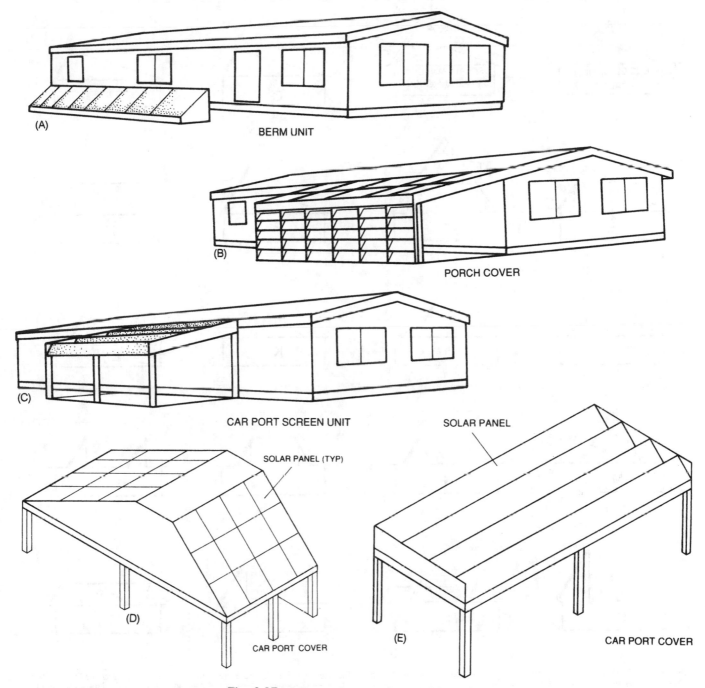

Fig. 8-17 *Collector placement for remodeling jobs.*

A computer service called SOLCOST provides a complete analysis based on the information supplied to the computer service. It includes a collector size optimization calculation that will provide the customer the optimum savings over the life of the equipment. For full details contact:

INTERNATIONAL BUSINESS SERVICES, INC.
Solar Group, 1010 Vermont Avenue
Washington, DC 20005

BUILDING MODIFICATIONS

In order to make housing more efficient, it is necessary to make some modifications in present-day carpentry practices. For instance, the following must be done to make room for better and more efficient insulation:

1. Truss rafters are modified to permit stacking of two 6-inch-thick batts of insulation over the wall plate. The truss is hipped by adding vertical members

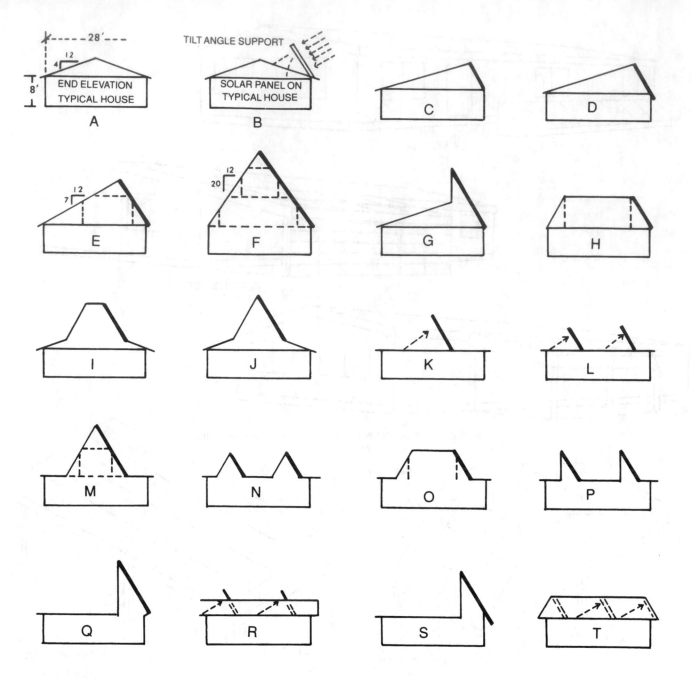

Elevations B, C, D, K, and L are probably the most economical methods of accommodating a steep tilt angle for solar collectors.

Elevations E, H, M, and O are probably the most economical.

Elevations I, J, M, and possibly O, depending on proportions, must be treated carefully.

Elevations G, N, P, Q, and S are possibilities but are more costly than the first group and especially more costly if clearstory space and/or fenestration are to be provided.

Elevations R and T are of some interest in that they provide a method of hiding the collectors completely, but their comparative cost needs special analysis.

Elevation F is obviously very expensive and only applicable under certain conditions. It does provide a three-story opportunity.

Fig. 8-18 *Collector roof configurations.* (National Association of Home Builders)

at each end and directly over the 2 × 6 studs of the outer wall.

2. All outside walls use thicker (2 × 6) studs on 2-foot centers to accommodate the thicker insulation batts.

3. Ductwork is framed into living space to reduce heat loss as warmed air passes through the ducts to the rooms.

4. Wiring is rerouted along the soleplate and through notches in the 2 × 6 studs. This leaves the insulation cavity in the wall free of obstructions.

5. Partitions join outside walls without creating a gap in the insulation. Drywall passes between the soleplates and abutting studs of the interior walls. The cavity is fully insulated.

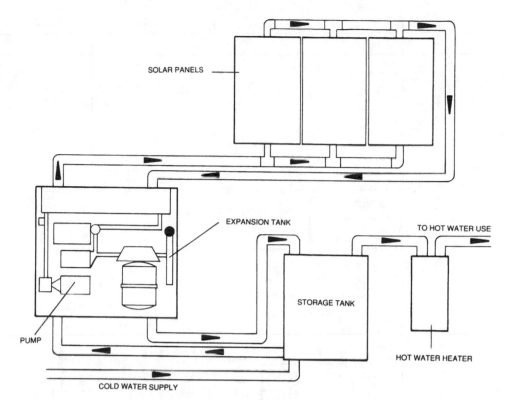

Fig. 8-19 *Prepackaged solar-assisted domestic hot water unit.*

6. Window area is reduced to 8 percent of the living area.

7. Box headers over the door and window openings receive insulation. In present practice, the space is filled with 2 × 10s or whatever is needed. Window header space can be filled with insulation. This reduces the heat loss through the wood that would normally be located here.

BUILDING UNDERGROUND

A number of methods are being researched for possible use in solar heating applications. One of the most inexpensive ways of obtaining insulation is building underground. However, one problem with underground living is psychological. People do not like to live where they can't see the sun or outside. The idea of building underground is not new. The Chinese have done it for centuries. But the problem comes in selling the public the idea. It will take a number of years of research and development before a move is made in this direction.

Advantages

There are a number of advantages of going underground. By using a subterranean design, the builder can take advantage of the earth's insulative properties.

The ground is slow to react to climatic temperature changes. It is a perfect year-round insulator. There is a relatively constant soil temperature at 30 feet below the surface. This could be ideal for moderate climates, since the temperature would be a constant 68 degrees F.

By building underground it would be possible to use the constant temperature to reduce heating and cooling costs. A substantial energy reduction or savings could be realized in the initial construction also.

Figure 8-21 shows a roof-suspended earth home. This is one of the designs being researched at Texas A&M University. It uses the earth as a building material, and uses the wind, water, vegetation, and the sun to modify the climate.

A suitable method of construction uses beams to support the walls. That means structural beams could be stretched across the hole and walls could be suspended from the beams. See Fig. 8-22. The inside and outside walls would also be dropped from the beams. That would allow something other than wood to be used for walls, since they would be non-weight-bearing partitions. The roof would be built several feet above the surface of the ground to allow natural lighting through the skylights and provide a view of outside. This would get rid of some of the feeling of living like a mole.

Another design being researched involves tunneling into the side of a hill. See Fig. 8-23. This still uses the insulative qualities of the earth and allows a southern exposure wall. This way windows and a conventional-type front door could be used. Exposing only the southern wall also helps to reduce heating costs. The steep hills in some areas could be used for this type

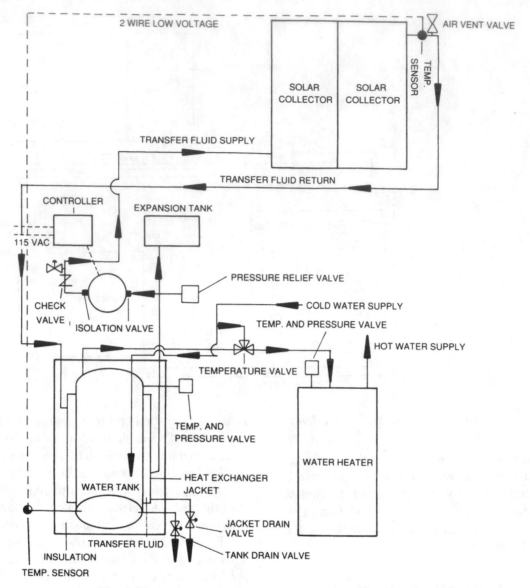

Fig. 8-20 *Piping and wiring diagram for solar water heater system.*

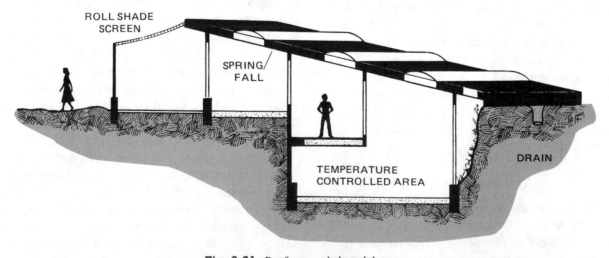

Fig. 8-21 *Roof-suspended earth home.*

Fig. 8-22 *Another variation of a roof-suspended earth home.*

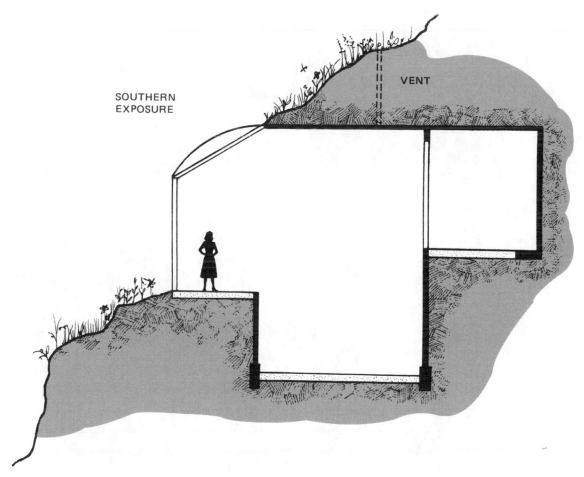

Fig. 8-23 *Hillside earth home.*

of housing where it would be impossible to build conventional-type housing.

Many more designs will be forthcoming as the need to conserve energy becomes more apparent. The carpenter will have to work with materials other than wood. The new methods and new materials will demand a carpenter willing to experiment and apply known skills to a rapidly changing field.

9
CHAPTER

Alternative Types of Foundations

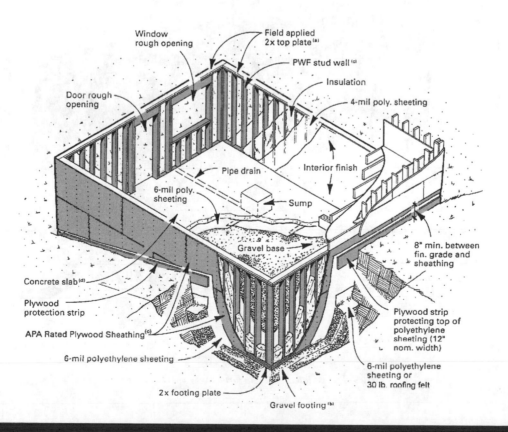

Window rough opening

Field applied 2x top plate [a]

PWF stud wall [c]

Insulation

4-mil poly. sheeting

Door rough opening

Pipe drain

Interior finish

6-mil poly. sheeting

Sump

Gravel base

8" min. between fin. grade and sheathing

Concrete slab [d]

Plywood protection strip

APA Rated Plywood Sheathing [c]

6-mil polyethylene sheeting

2x footing plate

Gravel footing [b]

Plywood strip protecting top of polyethylene sheeting (12" nom. width)

6-mil polyethylene sheeting or 30 lb. roofing felt

INSULATED CONCRETE FORMS

One of the fastest growing framing methods is the insulated concrete form (ICF). In this method, walls are made of concrete reinforced by rebar surrounded by Styrofoam® (expanded polystyrene beads). Most ICFs are hollow Styrofoam® blocks that are stacked together like building blocks with either tongue-and-groove joints or finger joints as shown in Fig. 9-1. Each manufacturer makes the block a different size. Some manufacturers have ties molded in their blocks while others have separate ties that are inserted into the block at the job site. Once the blocks are assembled, reinforced, and braced, concrete is poured into the blocks using a concrete pumper truck. An example of what the concrete and rebar would look like inside the foam is illustrated in Fig. 9-2.

Advantages and Disadvantages to Insulated Concrete Forms

Insulated concrete form homes are framed in concrete, which allows them to have the following advantages over conventional wood-frame homes:

- It can withstand hurricane force winds (200 mph).

- It is bullet resistant.

- Termites cannot harm its structural integrity.

- It exceeds building code requirements.

- It will lower heating and cooling costs by 50%–80%.

- Load bearing capacity is higher (27,000plf vs. 4,000plf).

- Outside noise reduction is higher.

- Homeowner's insurance rates should decrease 10%–25%.

- Fire rating of walls is measured in hours vs. minutes.

- Reduces pollen inside the home.

An ICF home is much stronger and more energy efficient than a wood-frame home. However, its major disadvantage is cost. The average ICF home costs about twice as much to frame as its chief competitor. Because the framing of a home is just one of its costs, this figure really amounts to only a 5%–15% increase in the total cost of a home.

Another disadvantage to an ICF is that there is no room for error. If a window or door opening is off, it is much more difficult to cut reinforced concrete than wood. Carpenters must also be retrained to use this framing method correctly.

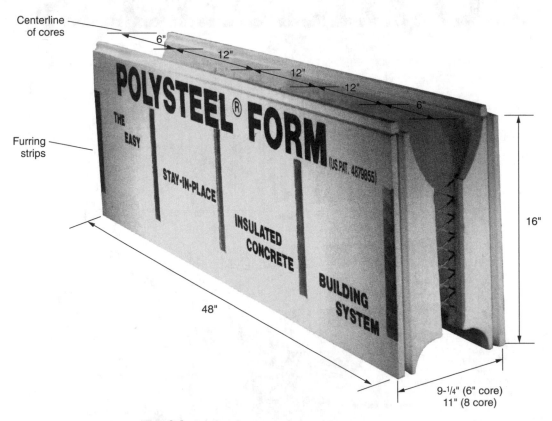

Fig. 9-1 *Insulated concrete form.* (American Polysteel Forms)

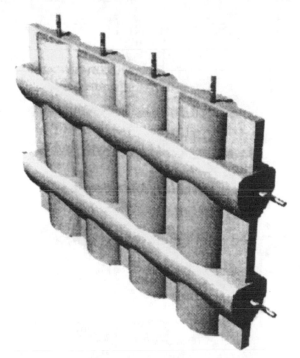

Fig. 9-2 *Concrete and rebar inside an insulated concrete form.*
(American Polysteel Forms)

Fig. 9-3 *Rebar inside the insulating plastic blocks.*

Fig. 9-4 *The plastic foam blocks can be easily fitted to the terrain.*

Tools Used in Insulated Concrete Form Framing

Because concrete can be molded into any shape, it is essential to mold the frame of a home straight and square so walls are plumb and level. In order to achieve this, a carpenter should have a framing square, 2- and 4-foot levels, chalk line, and 30- and 100-foot tape measures. The foam forms can easily be cut with a handsaw or sharp utility knife. In most cases, the foam forms are glued together, so a caulking gun is required.

All the foam forms need to be braced until the concrete hardens. If the construction crew is not using metal braces, then wooden braces need to be constructed. Therefore, a framing hammer, nails, circular saw, and crowbar are needed.

Rebar is run horizontally and vertically throughout the walls of an ICF home (Fig. 9-3). Tools used to work with rebar are rebar cutting and bending tools, metal cutoff saw, rebar twist tie tool (pigtail), 8-inch dikes, 9-inch lineman's pliers, hack saw, tin snips, and wire-cutting pliers. Figure 9-4 shows the placement of foam blocks cut to fit the terrain. A hot knife (Fig. 9-5) or router can be used to cut notches in the foam to place electrical wiring.

Sequence

Insulated concrete form framing is a radically new approach for carpenters that are familiar with the traditional wood "stick-built" method. The only wood used in this wall framing method is pressure-treated lumber, which is used for sealing rough door and window openings. Because this method is different and each manufacturer has specific guidelines that must be followed to ensure proper installation, training is offered to all individuals constructing ICFs. Many ICFs also send representatives to the job site to supervise those who are building their first home using this method. Therefore, only general guidelines are covered in the following section.

1. Before starting, rebar should extend vertically 2–6 feet from the foundation every one to two linear feet depending on building code requirements.

2. Most manufacturers suggest placing 2 × 4s or some type of bracing around the perimeter of the foundation as a guide for setting the foam forms.

3. Once the first course of foam form blocks is laid, continue to do so, placing rebar horizontally every one to two feet (or every other course of block) as required.

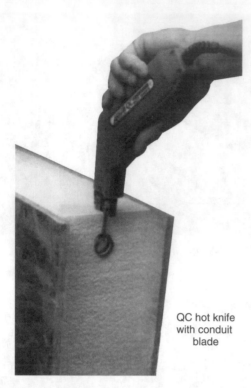

QC hot knife with conduit blade

Fig. 9-5 *A hot knife used to cut foam for electrical wiring and plumbing.* (Avalon Concepts)

4. Vertical bracing should be tied to the perimeter bracing at 6-foot intervals. Corners should be braced on each intersecting edge with additional diagonal bracing spaced at 4-foot intervals. See Fig. 9-6A, B, and C for a typical corner bracing diagram. Figure 9-6B shows how windows are blocked until the concrete solidifies. Figure 9-6C provides a closer look at a window blocked and braced. As forms are stacked, vertical bracing can be screwed into the metal or plastic ties built or inserted into the foam form blocks.

5. Openings for windows and doors should be blocked with pressure-treated lumber.

6. After the foam form blocks are stacked to the correct height, place bracing on the top and secure it to the side bracing. Because foam is very light, it will float. Proper bracing is essential to prevent the foam walls from floating and causing blowouts when the concrete is poured inside them. Additional bracing techniques are illustrated in Fig. 9-7A, B, C, and D.

7. Walls are usually poured with a boom pump truck in 4-foot increments. Anchor bolts are set in the top of the wall to secure the top plate as a nailing base for roof construction. See Fig. 9-8.

8. Sheathing can be screwed into the metal or plastic ties located every foot in the foam form block, as shown in Fig. 9-9.

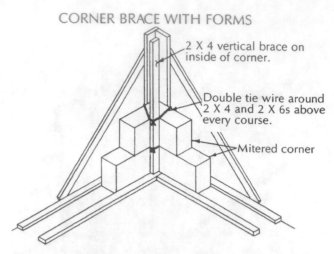

CORNER BRACE WITH FORMS

2 X 4 vertical brace on inside of corner.

Double tie wire around 2 X 4 and 2 X 6s above every course.

Mitered corner

Fig. 9-6A *Corner bracing technique for insulated concrete forms.*
(American Polysteel Forms)

Fig. 9-6B *Windows are blocked until after the concrete solidifies.*

Fig. 9-6C *Window blocked in and braced.*

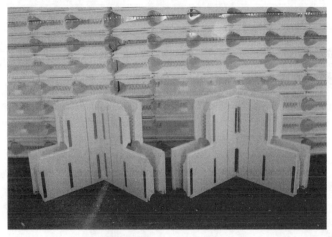

Fig. 9-7A *Corner insulated blocks.*

Fig. 9-7B *Bracing of corners and window forms.*

9. Electrical wiring and boxes can be installed by gouging out a groove with a router or hot knife. See Fig. 9-10 for electrical wiring tips.
10. A metal screen or mesh holds the foam block in place during the pouring operation. See Fig. 9-10A.

Types of Foam

The most commonly used types of foam utilized in ICFs are the *expanded polystyrene* (EPS), and *extruded polystyrene* (XPS). EPS consists of tightly fused beads of foam. Vending-machine coffee cups, for example, are made of EPS. XPS is produced in a different process and is more continuous, without beads or the sort of "grain" of EPS. The trays in prepackaged meat at the grocery store are made of XPS. The two types can differ in cost, strength, R-value, and water resistance. EPS varies somewhat. It comes in various densities; the most common are 1.5 pounds per cubic foot (pcf) and 2 pounds pcf. The denser foam is a little more expensive, but is a little stronger and has a slightly higher R-value.

Some stock EPS is now available with insect-repellent additives. Although few cases of insect penetration into the foam have been reported to date, some ICF manufacturers offer versions of their product made of treated material. Whether you buy your own foam or you are choosing a preassembled system, you might want to check with the manufacturer about this.

Three Types of ICF Systems

The main difference in ICF systems is that they vary in their unit sizes and connection methods. They can be divided into three types: *plank, panel,* and *block systems* as shown in Fig. 9-11. The panel system is the largest. It is usually 4 by 8 feet in size. That means the wall area can be erected in one step but may require more cutting. These panels have flat edges and are connected one to the other with fasteners such as glue, wire, or plastic channel.

Fig. 9-7C *Second-story bracing.*

Plank systems are usually 8 feet long with narrow (8- or 12-inch) planks of foam. These pieces of foam are held at a constant distance of separation by steel or plastic ties. The plank system has notched, cut, or drilled edges. The edges are where the ties fit. In addition to spacing the planks, the ties connect each course of planks to the one above and below.

Block systems include units ranging from the standard concrete block (8 × 16 inches) size to a much larger 16-inch-high by 4-foot-long unit. Along their edges are teeth or tongues and grooves for interlocking; they stack without separate fasteners on the same principle as children's Lego blocks.

Another difference is the shape of the cavities. Each system has one of three distinct cavity shapes. The shapes are flat, grid, or post-and-beam. These produce different shapes of concrete beneath the foam, as shown in Fig. 9-12.

Note how the flat cavities produce a concrete wall of constant thickness, just like a conventionally poured wall that was made with plywood or metal forms. Cavities are

2 X 4 TOP RAIL SUPPORT

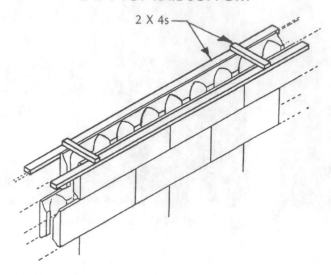

2 X 4s

HORIZONTAL TOP BRACE

◆ A horizontal top brace secured to the steel furring strips and supported by a diagonal brace staked to the ground will also provide the alignment and security necessary to keep the Forms in place during the pouring of concrete.

Fig. 9-7D *Additional bracing techniques.* (American Polysteel Forms)

POLYSTEEL FORM® TOP BRACE

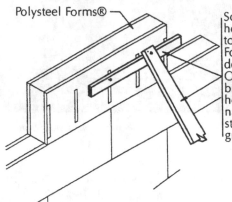

Polysteel Forms®

Screw a continuous horizontal 2 X 4 to the top course of Polysteel Form® wall with 2 1/2" deck screws at 4'–0" O.C., a 2 X 4 diagonal brace is screwed to the horizontal 2 X 4 and is nailed to a bracing stake driven into the ground.

Fig. 9-8 *Anchor bolt placement for roof framing.* (AFM)

Fig. 9-9 *Attaching sheathing directly to plastic ties in the concrete forms.* (AFM)

usually "wavy," both horizontally and vertically. If the forms are removed, it can be seen just how the walls resemble a breakfast waffle. The post-and-beam cavities are cavities with concrete only every few feet, horizontally and vertically. In the most extreme post-and-beam systems, there is a 6-inch-diameter concrete "post" formed every 4 feet and a 6-inch concrete "beam" at the top of each story.

Keep in mind that no matter what the shape of the cavity, all systems have "ties." These are the crosspieces that connect the front and back layers of foam that make up the form. If the ties are plastic or metal, the concrete is not affected significantly. However, in some grid systems, they are foam and are much larger. But the forming breaks in the concrete about 2 inches in diameter every foot or so. See Fig. 9-12 for the differences.

Fig. 9-10A *Placement of a metal screen or mesh to hold the foam block during the concrete pouring operation.*

Table 9-1 shows eight different systems used in ICF with their dimensions, fastening surface, and various notes with additional details. Note that no matter what the cavity shape is, all systems also have "ties." These ties are the crosspieces that connect the front and back layers of foam. When the ties are metal or plastic, they do not affect the shape of the concrete much. But in some of the grid systems, they are foam and are much larger, forming breaks in the concrete about 2 inches in diameter every foot or so. Figure 9-13 shows the differences.

Another difference is that many of the systems also have a fastening surface which is some material other than foam, embedded into the units that crews can sink screws or nails into the same way as fastening to a stud. Often this surface is simply the ends of the ties; however, other systems have no embedded fastening surface. These units are all foam, including the ties. This generally makes them simpler and less expensive but requires crews to take extra steps to connect interior wallboard, trim, exterior siding, and so on to the walls.

Cutaway views of the wall panels are shown in Figs. 9-14 through 9-21. In Fig. 9-22 the R-Forms panel is being assembled on site. In Fig. 9-23, there is an on-site pile of ENER-GRID panels. Figure 9-24 shows the top view of an Amhome panel that has furring strips embedded. Note how light the panels are. They are easily handled by one person. The worker is carrying a fold-out Fold-Form block before and after it has been spread for adding to the construction (Fig. 9-25).

Fig. 9-10B *Installing electrical wiring and boxes in insulated concrete forms.* (American Polysteel Forms)

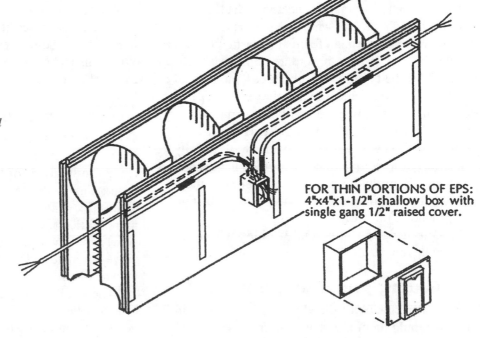

FOR THIN PORTIONS OF EPS:
4"x4"x1-1/2" shallow box with single gang 1/2" raised cover.

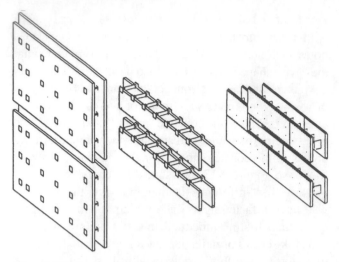

Fig. 9-11 *Diagrams of ICF formwork made with the three basic units: panel on the left, plank in the center, and block on the right.* (PCC)

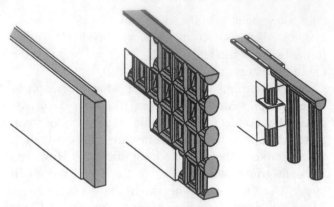

Fig. 9-12 *Cutaway diagrams of ICF walls with the three basic cavity shapes: flat, grid, and post-and-beam.* (PCC)

Two varieties of SmartBlock are shown in Fig. 9-26. Three of them, A, C, and D, are assembled with plastic ties. A grid block without fastening surfaces is shown in Fig. 9-26B.

Figure 9-27A, B, C, and D shows setting the insulating concrete forms, how the completed forms look, pumping in concrete, and the completed house.

Foam Working Tools

Some of the tools needed for working with foam are not familiar to the usual framer of houses. These tools may include the thermal cutter. This cutter is a new tool that cuts a near-perfect line through foam and plastic units in one pass. Figure 9-28 shows this device. It is made up of a taut resistance-controlled wire mounted on a bench-frame. It is heated with electricity and drawn though the unit while the wire is red hot. It melts a narrow path through foam and plastic ties. It is a worthwhile tool to have if you are building a high-volume ICF walled house. It will not cut rough metal ties or the foam-and-cement material of the grid panel systems. However, companies selling thermal cutters are usually located in every community. The grid panel systems can be cut with any of the bladed tools used by a carpenter, but sometimes a chain saw is a handy tool. It goes quickly through the heavier material of these systems and cuts through in one pass, whether cutting on the ground or in place.

Tools needed for cutting and working with foam are shown in Table 9-2.

Gluing and Tying Units

ICF units are frequently glued at the joints to hold them down, hold them together, and prevent concrete leakage. Common wood glue and most construction adhesives do the job well. Popular brands are Liquid Nail, PL 200, and PL 400. Some of these can dissolve the foam but, if applied in a thin layer, the amount of foam lost is usually insignificant. Look for an adhesive that is "compatible with polystyrene." Figure 9-29 shows an industrial foam gun.

Rebar is often precut to length and pre-bent but, even if it is, the workers generally have to process a few bars in the field. Most ICF systems also have cradles that hold the bars in place for the pour, but a few bars need to be wired to one another or to ties to keep them in position.

It is possible to bend rebar with whatever tools are handy and cut it with a hacksaw; however, if you process large quantities, you might prefer buying or renting a cutter-bender. A large manual tool is pictured in Fig. 9-30. It makes the job faster and easier. Cutters-benders are available at steel-supply, concrete-supply, and masonry-supply houses. Almost any steel wire can hold rebar in place, but most efficient are rolls of precut wires shown in Fig. 9-31. These wires make the job faster and easier. Wire coils and belt-mounted coil holders are both sold by suppliers of concrete products, masonry, and steel.

Pouring Concrete

Concrete is best poured at a more controlled rate into ICFs than it is into conventional forms. An ordinary chute can be used for foundation walls (basement or stem); this is the least expensive option because it comes free with the concrete truck. Precise control is more difficult with a chute. You must pour more slowly than with conventional forms, and you must move the chute and truck frequently to avoid overloading any one section of the formwork.

The smaller line pump pushes concrete through a hose that lies on the ground. See Fig. 9-32. The crew holds the end of the hose over the formwork to drop concrete inside. If possible, use a 2-inch hose. One or two workers can handle it, and it can generally be run at full speed without danger. If only a 3-inch hose is available, you can use it; but pump slowly until you learn how much pressure the forms can take.

Boom pumps are mounted on a truck that also holds a pneumatically operated arm (the boom). The hose from the pump causes the concrete to move along the length of the boom and then hang loose from the end. See Fig. 9-33. By moving the boom, the truck's operator can dan-

Table 9-1 *Available ICF Systems*[1] *(Portland Cement Assoc.)*

	Dimensions[2] (width x height x length)	Fastening surface	Notes
Panel systems			
Flat panel systems			
R-FORMS	8" x 4' x 8'	Ends of plastic ties	Assembled in the field; different lengths of ties available to form different panel widths.
Styroform	10" x 2' x 8'	Ends of plastic ties	Shipped flat and folded out in the field; can be purchased in larger/smaller heights and lengths.
Grid panel systems			
ENER-GRID	10" x 1'3" x 10'	None	Other dimensions also available; units made of foam/cement mixture.
RASTRA	10" x 1'3" x 10'	None	Other dimensions also available; units made of foam/cement mixture.
Post-and-beam panel systems			
Amhome	9 3/8" x 4' x 8'	Wooden strips	Assembled by the contractor from foam sheet. Includes provisions to mount wooden furring strips into the foam as a fastening surface.
Plank systems			
Flat plank systems			
Diamond Snap-Form	1' x 1' x 8'	Ends of plastic ties	
Lite-Form	1' x 8" x 8'	Ends of plastic ties	
Polycrete	11" x 1' x 8'	Plastic strips	
QUAD-LOCK	8" x 1' x 4'	Ends of plastic ties	
Block systems			
Flat block systems			
AAB	11.5" x 16⅝" x 4'	Ends of plastic ties	
Fold-Form	1' x 1' x 4'	Ends of plastic ties	Shipped flat and folded out in the field.
GREENBLOCK	10" x 10" x 3'4"	Ends of plastic ties	
SmartBlock Variable Width Form	10" x 10" x 3'4"	Ends of plastic ties	Ties inserted by the contractor; different length ties available to form different block widths.
Grid block systems with fastening surfaces			
I.C.E. Block	9 1/4" x 1'4" x 4'	Ends of steel ties	
Polysteel	9 1/4" x 1'4" x 4'	Ends of steel ties	
REWARD	9 1/4" x 1'4" x 4'	Ends of plastic ties	
Therm-O-Wall	9 1/4" x 1'4" x 4'	Ends of plastic ties	
Grid block systems without fastening surfaces			
Reddi-Form	9 5/8" x 1' x 4'	Optional	Plastic fastening surface strips available
SmartBlock Standard Form	10" x 10" x 3'4"	None	
Post-and-beam block systems			
ENERGYLOCK	8" x 8" x 2'8"	None	
Featherlite	8" x 8" x 1'4"	None	
KEEVA	8" x 1' x 4'	None	

[1] All systems are listed by brand name.
[2] "Width" is the distance between the inside and outside surfaces of foam of the unit. The thickness of the concrete inside will be less, and the thickness of the completed wall with finishes added will be greater.

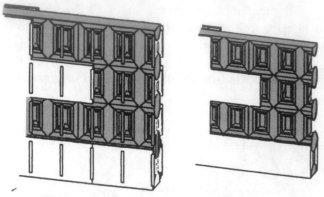

Fig. 9-13 *Cutaway diagrams of ICF grid walls with steel/plastic ties and foam ties.* (PCC)

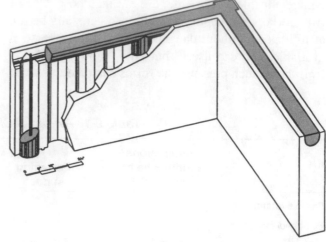

Fig. 9-16 *Cutaway diagram of a post-and-beam panel wall.* (PCC)

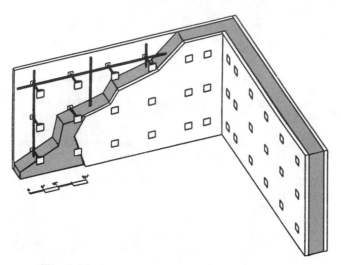

Fig. 9-14 *Cutaway diagram of a flat panel wall.* (PCC)

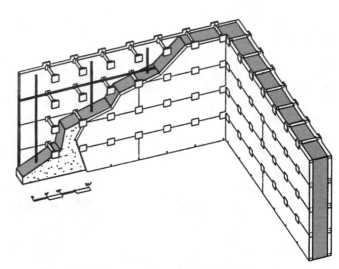

Fig. 9-17 *Cutaway diagram of a flat plank wall.* (PCC)

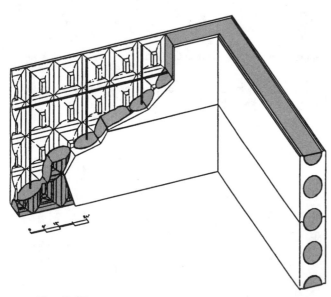

Fig. 9-15 *Cutaway diagram of a grid panel wall.* (PCC)

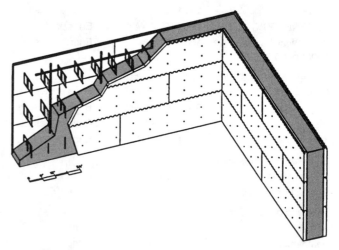

Fig. 9-18 *Cutaway diagram of a flat block wall.* (PCC)

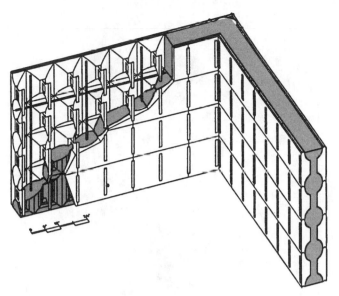

Fig. 9-19 *Cutaway diagram of a grid block wall with fastening surfaces. (PCC)*

Fig. 9-20 *Cutaway diagram of a grid block wall without fastening surfaces. (PCC)*

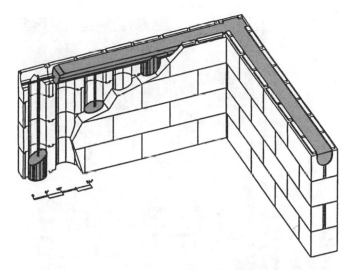

Fig. 9-21 *Cutaway diagram of a post-and-beam block wall. (PCC)*

Fig. 9-22 *Site-assembly of R-Forms panel. (PCC)*

gle the hose wherever the crew calls for it. One worker holds the free end to position it over the form-work cavities. The standard hose diameter is 4 inches. See Figs. 9-34, 35, 36.

You will need to have the hose diameter reduced to 2 or 3 inches with tapered steel tubes called "reducers." The narrower diameter slows down and smoothes out the flow of the concrete. Figure 9-34 shows how the reducers are arranged on a boom with hoses. Also, ask for two 90-degree elbow fittings on the end of the hose assembly, as seen in Fig. 9-34. These form an "S" in the line that further breaks the fall of the concrete. The concrete can be rough leveled off on top of the form by using a mason's

trowel. Figures 9-35–9-40 show pouring concrete and the finished project.

CONCRETE BLOCK

The second most popular residential framing method is concrete block. Comprised of less than 5% of the residential framing market, concrete block is another traditional method that has been around for a long time. It is mainly used when home owners are interested in a framing method that is resistant to inclement weather and termites. This type of construction is much more costly and time-consuming. Because the techniques associated with masonry and concrete work are numerous and detailed, only the major steps are reviewed.

1. Snap a chalk line on the footing or foundation where the concrete block should be placed.

Fig. 9-23 *An ENER-GRID panel.* (PCC)

(a)

Fig. 9-24 *Top view of an Amhome panel with embedded furring strips.* (PCC)

2. Spread mortar an inch or two thick on the footing or foundation and an inch or so wider and longer than the block.

3. Place a concrete block on each corner of the wall. Make sure both blocks are level and straight.

4. Stretch a string line tightly between the front, top edge of both blocks. This can be done by wrapping a string around a brick and placing it on top of the concrete block.

(b)

Fig. 9-25 *Folding out a Fold-Form block before use.* (PCC)

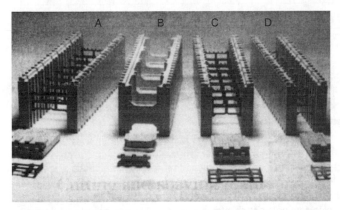

Fig. 9-26 *The two varieties of SmartBlock: a flat block assembled with plastic ties (A,C,D) and a grid block without fastening surfaces in B.* (PCC)

5. Place concrete blocks evenly about seven blocks (if using 8" × 16" block) from the corner. Do this for both corners.

6. Continue to build up the corners by backing up each level of block (course) by half a block. When finished, one block should be on the top of each corner. Each corner is called a *lead*.

Fig. 9-28 *A thermal cutter. Note the white line that is part of the cutting device. It is attached to the transformer on the left of the framework.* (PCC)

7. Place concrete blocks between each lead using a string to keep the top and front edge of the block straight and level. Mortar should be placed on one

A

B

C

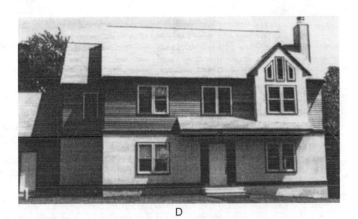

D

Fig. 9-27 *A. Setting insulating concrete forms. B. Completed formwork. C. Pumping in the concrete. D. The completed home.* (PCC)

Table 9-2 *Useful Tools and Materials* (Portland Cement Assoc.)

Operation or Class of Material	
Tool or Material	**Comments**
Cutting and shaving foam	
Drywall or keyhole saw	For small cuts, holes, and curved cuts.
PVC or mitre saw	For small, straight cuts and shaving edges.
Coarse sandpaper or rasp	For shaving edges.
Bow saw or garden pruner	For faster straight cuts.
Circular saw	For fast, precise, straight cuts. For cutting units with steel ties, reverse the blade or use a metal-cutting blade.
Reciprocating saw	For fast cuts, especially in place.
Thermal cutter	For fast, very precise cuts on a bench. Not suitable for steel ties or grid panel units.
Chain saw	For fast cuts of grid panel units.
Lifting units	
Forklift, manual lift, or boom or crane truck	For carrying large grid panel units and setting them in place. For upper stories, a truck is necessary.
Gluing and tying units	
Wood glue, construction adhesive, or adhesive foam	
Small-gage wire	For connecting units of flat panel systems.
Bending, cutting, and wiring rebar	
Cutter-bender	
Small-gage wire or precut tie wire or wire spool	
Filling and sealing formwork	
Adhesive foam	
Placing concrete	
Chute	For below-grade pours.
Line pump	Use a 2-inch hose.
Boom pump	Use two "S" couplings and reduce the hose down to a 2-inch diameter.
Evening concrete	
Mason's trowel	
Dampproofing walls below grade	
Nonsolvent-based dampproofer or nonheat-sealed membrane product	
Surface cutting foam	
Utility knife, router, or hot knife	Heavier utility knives work better. Use a router with a half-inch drive for deep cutting.
Fastening to the wall	
Galvanized nails, ringed nails, and drywall screws	For attaching items to fastening surfaces. Use screws only for steel fastening surfaces.
Adhesives	For light and medium connections to foam.
Insulation nails and screws	For holding lumber inside formwork.
J-bolt or steel strap	For heavy structural connections.
Duplex nails	For medium connections to lumber.
Small-gage wire	For connecting to steel mesh for stucco.
Concrete nails or screw anchors	For medium connections to lumber after the pour.
Flattening foam	
Coarse sandpaper or rasp	For removing small high spots.
Thermal cutter	For removing large bulges.
Foam	
Expanded polystyrene or extruded polystyrene	Consider foam with insect-repellent additives
Concrete	
Midrange plasticizer or superplasticizer	For increasing the flow of concrete without decreasing its strength. Can also be accomplished by changing proportions of the other ingredients.
Stucco	
Portland cement stucco or polymer-based stucco	

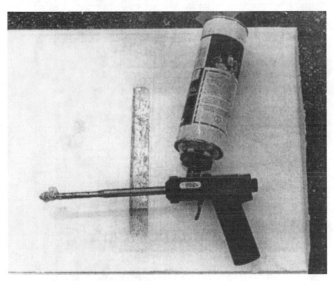

Fig. 9-29 *The industrial foam gun can be used as a glue gun.* (PCC)

Fig. 9-30 *This is a rebar cutter-bender.* (PCC)

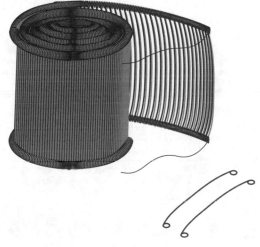

Fig. 9-31 *Roll of precut tie wires.* (PCC)

end of each block. A ⅜-inch-thick layer of mortar is common, because the actual size of most 8" × 16" concrete block is 7⅝" × 15⅝".

8. Allowances should be made for rough openings of windows and doors. Lintels or metal ledges are used to support block over door and window openings.

9. The top course of concrete block is filled with mortar, and anchor bolts are inserted at certain intervals according to building code requirements.

10. Mortar joints are usually smoothed (tooled) with a round bar, square bar, or s-shaped tool.

THE PERMANENT WOOD FOUNDATION SYSTEM

The Permanent Wood Foundation (PWF) is an innovative building system that saves builders time. It creates comfortable, warm living areas that enhance a home's value. A PWF system consists of a load-bearing frame made of pressure-treated lumber and sheathed with

Fig. 9-32 *Line pump for concrete.* (PCC)

Fig. 9-33 *Boom pump for pouring concrete.* (PCC)

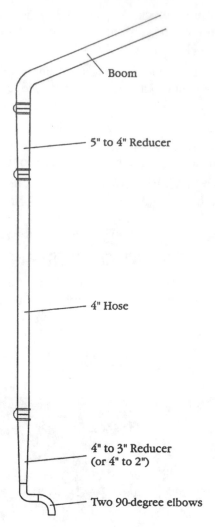

Boom

5" to 4" Reducer

4" Hose

4" to 3" Reducer
(or 4" to 2")

Two 90-degree elbows

Fig. 9-34 *Diagram of boom pump fittings that are suitable for pouring ICF walls. (PCC)*

Fig. 9-35 *Pouring concrete into the forms using a concrete mixer with a pump.*

Fig. 9-36 *Pouring in various weather conditions is possible since the foam form insulates the concrete as it sets.*

Fig. 9-37 *Note how the concrete is already in the form with the steel rods permanently set in place*

pressure-treated plywood. This type of construction is suitable for crawl spaces as well as for split entry or full basements. See Fig. 9-41.

Builders and homeowners have chosen the PWF system for good reasons, including design flexibility, faster construction, and larger, more comfortable living space.

The International Building Code recognizes both the one- and two-family dwelling with a PWF foundation. The Code (ICC), as well as many local and national organizations, recognizes this type of construction.

This system is a proven one that has been demonstrated by long-term ground tests conducted by the USDA's Forest Service. In these tests, pressure-treated wood withstood severe decay and termite conditions during decades of exposure. The walls are designed to resist and distribute the earth, wind, and seismic loads and stresses that may crack other types of foundations. The Permanent Wood Foundation is accepted by the major building codes, by federal agencies, and by

Fig. 9-38 Framing and bracing for the garage doorway.

Fig. 9-39 Second story forms are in place and braced.

Fig. 9-40 The finished project with a stucco finish on the exterior of the house.

lending, home warranty, and fire insurance institutions. This building system has been proven by years of success in more than 300,000 homes and other structures throughout the United States. Like conventional wood frame walls, the wood foundation is adaptable to virtually any design. The PWF fits a variety of floor plans. It can be used for both level and sloping sites. See Fig. 9-42.

Virtually any home design—traditional or contemporary—can benefit from the many advantages of a PWF. Room additions are also simplified.

Panel Foundations

Panel foundations can be site built or made in a shop. Third party inspection may be required; look for grade stamps, treatment stamps, stud spacing, insulation, nailing, depth of saw cuts, and plywood requirements, which are some of the items that a third party inspection will note. All panel wood foundations must be designed and installed in accordance with current Building Code Standards.

Wood foundations are easy to build if one is building from an accurately designed plan. When a plan is incorrect, or if something is left out of the plan, or if a design is made using a guide manual, major mistakes can be made during the construction process. Keep in mind that a guide manual is not a design manual. These errors cause problems for the owner, builder, and the building department.

The wood used in the wood foundation should be waterproofed by using chemically saturated wood to prevent rot and attack by various insects, termites being the most bothersome. Water-borne preservatives are applied to the plywood and lumber in a high-pressure chamber where the preservatives are driven deep into the wood fiber and permanently bonded in the cellular structure of the wood. After treatment, the wood is dried to normal moisture levels. Pressure-preserved wood retains all its stiffness, strength, and workability, and has the added value of permanent protection against mold, decay, and insect infestation.

Building Materials

High-quality building materials are used for the Permanent Wood Foundation. Plywood must meet the strict requirements of U.S. Product Standard PS 1-83 for construction and industrial plywood. Lumber must be of a species for which allowable unit stresses are given in the "National Design Specification for Wood Construction" and must be grade-marked by an approved inspection agency. Both plywood and lumber must be pressure-treated in accordance with the penetration, retention, and drying requirements of the American Wood Preservers Bureau AWPB-FDN Standard. Required preservative retention is 50 percent higher than code requirements for normal ground contact application.

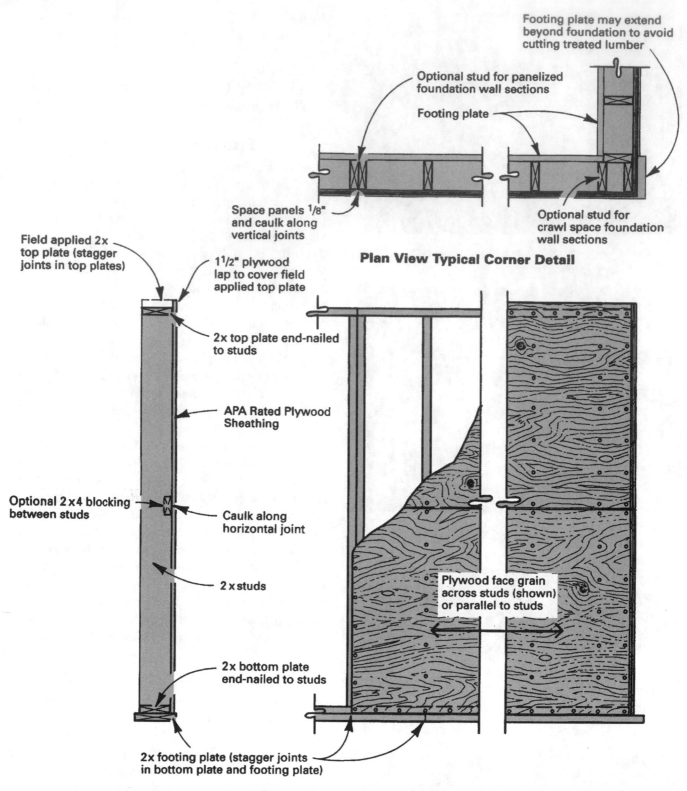

Plan View Typical Corner Detail

Footing plate may extend beyond foundation to avoid cutting treated lumber

Optional stud for panelized foundation wall sections

Footing plate

Optional stud for crawl space foundation wall sections

Space panels 1/8" and caulk along vertical joints

Field applied 2x top plate (stagger joints in top plates)

1½" plywood lap to cover field applied top plate

2x top plate end-nailed to studs

APA Rated Plywood Sheathing

Optional 2x4 blocking between studs

Caulk along horizontal joint

2x studs

Plywood face grain across studs (shown) or parallel to studs

2x bottom plate end-nailed to studs

2x footing plate (stagger joints in bottom plate and footing plate)

Fig. 9-41A *Foundation panel construction details.*

The Permanent Wood Foundation is the result of extensive design and engineering analyses by the U.S. Department of Agriculture Forest Service, the National Forest Products Association, and the National Association of Home Builders Research Foundation.

Additional in-ground structural testing has been conducted by the American Plywood Association. As mentioned above, durability of the system was demonstrated in long-term tests conducted by the Forest Service. Pressure-preservative-treated wood in these tests

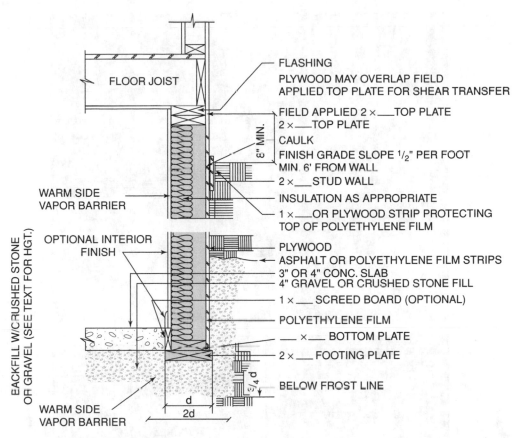

FLASHING

PLYWOOD MAY OVERLAP FIELD APPLIED TOP PLATE FOR SHEAR TRANSFER

FIELD APPLIED 2 × ___TOP PLATE
2 × ___TOP PLATE
CAULK
FINISH GRADE SLOPE 1/2" PER FOOT
MIN. 6' FROM WALL
2 × ___STUD WALL
INSULATION AS APPROPRIATE
1 × ___OR PLYWOOD STRIP PROTECTING TOP OF POLYETHYLENE FILM

PLYWOOD
ASPHALT OR POLYETHYLENE FILM STRIPS
3" OR 4" CONC. SLAB
4" GRAVEL OR CRUSHED STONE FILL
1 × ___ SCREED BOARD (OPTIONAL)
POLYETHYLENE FILM
___ × ___ BOTTOM PLATE
2 × ___ FOOTING PLATE
BELOW FROST LINE

FLOOR JOIST

WARM SIDE VAPOR BARRIER

8" MIN.

OPTIONAL INTERIOR FINISH

BACKFILL W/CRUSHED STONE OR GRAVEL (SEE TEXT FOR HGT.)

WARM SIDE VAPOR BARRIER

d
2d
3/4 d

Fig. 9-41B *How the basement wall is placed in reference to the base and floor joist.*

withstood severe decay and termite conditions with treatment levels lower than required by the AWPB standard for Permanent Wood Foundations. As a result, the PWF basement has all the comforts of an above ground room. Real wood construction lends a feeling of warmth—not the musty, damp feeling usually associated with basements. The wood basement is an integral part of the home. This type of basement allows for every square foot to become living space—below ground and above.

Energy Considerations

The wood framing of the Permanent Wood Foundation makes it easy to install thick, economical batt-type insulation. This means less heat loss through the walls and greater long-term energy savings. The cost of installing insulation is less, too, because the wood frame walls are already in place. Consider this comparison: In a concrete foundation, ¾- or 1½-inch foam sheathing is typically installed for insulation. This gives an energy rating of R-3 to R-6. In the wood foundation, it's easy to fit 3½-inch-thick insulation between the 2 × 4 studs, producing an R-11 energy rating. If 5½-inch insulation is installed between 2 × 6 studs, the energy rating increases to R-19.

Finishing

The Permanent Wood Foundation is easier to finish inside than conventional foundations because the wooden studs are already in place. Plumbing and wiring are simplified and concealed. No furring strips are needed for installing insulation, gypsum board, or paneling. Because it's so easy, many PWF homeowners do the finishing themselves. That way, it can be done according to their own tastes, imagination, and at less cost.

Adding Living Space

PWFs provide more actual living space in basements than concrete or masonry basements of the same dimensions. That's because the wood foundation walls don't need to be as thick as concrete or masonry walls. Insulation takes less space in the PWF because it fits into the cavities of the wood frame wall. That means extra furring strips aren't necessary. This, in turn, produces more living area. See Fig. 9-43.

Remodeling

The PWFs are easy to remodel or modify. You can cut out window or door openings or add a whole room. Simple finishing touches such as pictures and shelves

Fig. 9-41C *Applying glue to the floor joists before laying the plywood subfloor.*

can be hung without special tools or fasteners. When adding on to a home, you won't have to worry about providing access for a concrete truck. Permanent Wood Foundations have even been used to replace old and weakened mason and concrete foundations—with considerable savings in time and expense.

Flexibility

Flexibility is one of the primary considerations when selecting a Permanent Wood Foundation. The system can be used in both single- and multistory structures and for both site-built and manufactured homes. The system can be adapted to almost any home design and site plan. It can be engineered for a variety of soil conditions, even those with high water tables. It is as easy to install for crawl space designs as for split-entry and full basement buildings.

PRESSURE-TREATED WOOD CONCERNS AND CONSIDERATIONS

When most wood is exposed to the elements, excessive moisture, or contact with the ground, it will decompose. That is because four conditions are required for decay and insect attack to occur:

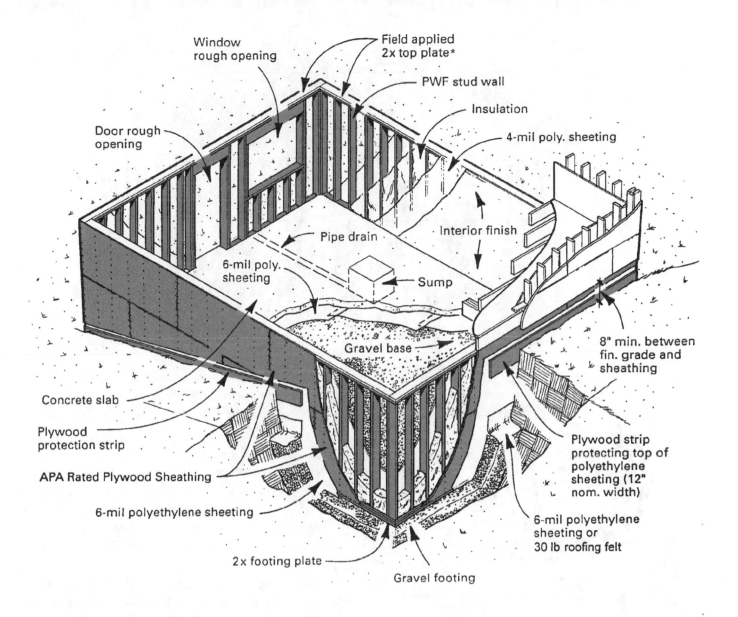

Window rough opening

Field applied 2x top plate*

PWF stud wall

Insulation

4-mil poly. sheeting

Door rough opening

Pipe drain

Interior finish

6-mil poly. sheeting

Sump

8" min. between fin. grade and sheathing

Gravel base

Concrete slab

Plywood protection strip

Plywood strip protecting top of polyethylene sheeting (12" nom. width)

APA Rated Plywood Sheathing

6-mil polyethylene sheeting

6-mil polyethylene sheeting or 30 lb roofing felt

2x footing plate

Gravel footing

* Not required to be treated if exterior finish grade is 8″ or more below bottom of plate.

Fig. 9-42 *This is a typical wood foundation basement.*

- Moisture
- Temperature of approximately 50 to 90°F
- Oxygen
- A source of food (wood fiber)

If any of these conditions is removed, infestation and decomposition will not occur.

Southern pine has long been a preferred species when pressure treatment with preservatives is required. This is because of the ease with which it can be treated. The unique cellular structure of southern pine permits deep, uniform penetration of preservative chemicals, rendering the wood useless as a food source for fungi,

termites, and microorganisms. Some 85 percent of all pressure-treated wood is southern pine.

Why design and build with pressure-treated wood? Figure 9-44 shows, by zones, the level of wood deterioration throughout the United States. As shown, deterioration zones from moderate to severe cover most of the country.

Modern science has developed preservative treatments that are odorless and colorless and leave the wood preserved and dry to the touch. Treatment with chemical preservatives protects wood that is exposed to the elements, is in contact with the ground, or is used in high humidity areas.

Fig. 9-43 *A cutaway showing the comfortable room addition possible by using a PWF for the house.*

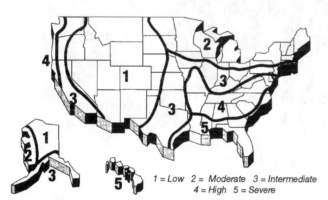

1 = Low 2 = Moderate 3 = Intermediate
4 = High 5 = Severe

Fig. 9-44 *Wood deterioration zones in the United States.*

However, beware—not all treated wood is equal. Most wood species do not readily accept chemical preservatives and must first be "incised" or perforated by a series of small slits along the grain of the wood. Incising allows sufficient penetration of the preservatives to meet American Wood Preservers Association standards. Southern pine doesn't require incising. See Table 9-3 for softwood lumber and plywood standards.

In addition, the use of treated southern pine provides no measurable risk to humans, animals, plants, or other life. Scientific research studies have shown the following:

• Preservative-treated wood products last longer than alternative materials, conserving a renewable natural resource.

• Wood preservatives do not aggressively leach into the ground, waterways, or drinking water, or adversely affect marine life.

• Preservative-treated wood products have been extensively tested and proven to be more durable than alternative products that require more energy to produce and that may also be aesthetically unacceptable to consumers.

How long does pressure-treated wood last? Currently available research shows that wood that has been properly treated and installed for its intended use can be expected to last for many decades.

Ongoing tests are sponsored and monitored by the USDA Forest Service's Forest Products Laboratory. Test stakes of treated wood have been buried in the ground at various locations, stretching from the Mississippi Delta to the Canadian border. Data analysis indicates that CCA-treated southern pine stakes, still in place since 1938, have shown no failures at chemical retention levels of 0.29 pounds of preservatives per foot of wood or higher.

TYPES OF WOOD PRESERVATIVE

Several waterborne preservatives are commonly used, including

• Chromated copper arsenate (CCA)

• Alkaline copper quat (ACQ)

Table 9-3 *AWPA Standards for Softwood Lumber & Plywood*

	USE CATEGORY STANDARDS				COMMODITY STANDARDS

SELECTION GUIDE

Use Category	Service Conditions	Use Environment	Common Agents of Deterioration	Typical Applications
UC1	Interior construction Above Ground Dry	Continuously protected from weather or other sources of moisture	Insects only	Interior construction and furnishings
UC2	Interior construction Above Ground Damp	Protected from weather, but may be subject to sources of moisture	Decay fungi and insects	Interior construction
UC3A	Exterior construction Above Ground Coated & rapid water runoff	Exposed to all weather cycles, not exposed to prolonged wetting	Decay fungi and insects	Coated millwork, siding and trim
UC3B	Exterior construction Above Ground Uncoated or poor water runoff	Exposed to all weather cycles including prolonged wetting	Decay fungi and insects	Decking, deck joists, railings, fence pickets, sill plates, uncoated millwork
UC4A	Ground contact or fresh water Non-critical components	Exposed to all weather cycles, normal exposure conditions	Decay fungi and insects	Fence, deck, and guardrail posts, crossties & utility poles (low decay areas)
UC4B	Ground contact or fresh water Critical components or difficult replacement	Exposed to all weather cycles, high decay potential including salt water splash	Decay fungi and insects with increased potential for biodeterioration	Permanent wood foundations, building poles, horticultural posts, crossties & utility poles (high decay areas)
UC4C	Ground contact or fresh water Critical structural components	Exposed to all weather cycles, severe environments, extreme decay potential	Decay fungi and insects with extreme potential for biodeterioration	Land & fresh water piling, foundation piling, crossties & utility poles (severe decay areas)
UC5A	Salt or brackish water and adjacent mud zone Northern waters	Continuous marine exposure (salt water)	Salt water organisms	Piling, bulkheads, bracing
UC5B	Salt or brackish water and adjacent mud zone New Jersey to Georgia, South of San Francisco	Continuous marine exposure (salt water)	Salt water organisms including Limnoria tripunctata	Piling, bulkheads, bracing
UC5C	Salt or brackish water and adjacent mud zone South of Georgia and Gulf Coast	Continuous marine exposure (salt water)	Salt water organisms including Martesia, Sphaeroma	Piling, bulkheads, bracing
UCFA	Fire protection as required by codes Above Ground Interior construction	Continuously protected from weather or other sources of moisture	Fire	Roof sheathing, roof trusses, studs, joists, paneling
UCFB	Fire protection as required by codes Above Ground Exterior construction	Subject to wetting	Fire	Vertical exterior walls, inclined roof surfaces or other construction which allows water to quickly drain

COMMODITY STANDARDS

C1 – All Timber Products – Preservative Treatment by Pressure Processes

C2 – Lumber, Timber, Bridge Ties and Mine Ties

C3 – Piles

C4 – Poles

C5 – Fence Posts

C6 – Crossties and Switch Ties

C9 – Plywood

C11 – Wood Blocks for Floors and Platforms

C14 – Wood for Highway Construction

C15 – Wood for Commercial – Residential Construction

C16 – Wood Used on Farms

C17 – Playground Equipment

C18 – Marine Construction

C20 – Structural Lumber: Fire Retardant Treatment by Pressure Processes

C22 – Permanent Wood Foundations

C23 – Round Poles and Posts used in Building Construction

C24 – Sawn Timber used to Support Residential & Commercial Structures

C25 – Sawn Crossarms

C27 – Plywood Fire Retardant Treatment by Pressure Processes

C28 – Glued-Laminated Members

C29 – Lumber and Plywood to be used for the Harvesting, Storage and Transportation of Food Stuffs

C30 – Lumber, Timbers and Plywood for CoolingTowers

C31 – Lumber used Out of Contact with the Ground and Continuously Protected from Liquid Water

C32 – Glue Laminated Poles

C33 – Structural Composite Lumber

C34 – Shakes & Shingles

NOTE: The Standards are not in consecutive order, as some have been deleted over time. Only those pertaining to Southern Pine have been included here.

SPECIAL NOTE: The American Wood-Preservers' Association (AWPA) developed the Use Category System (UCS) to provide a simple way to use AWPA Standards. The UCS defines a series of different exposures for treated wood products. Each exposure has a different degree of biodegradation hazard and/or product service life expectation. All treated wood commodities can be placed into one of the Use Categories (see table above). There are five Use Categories based on exposures and expected product performance, ranging from weather protected to salt water marine. A separate Use Category is provided for fire retardant applications. The Use Category number relates to the hazard associated with certain use environments, while the letter following the number (if present) relates to the risk. In general, as the Use Category number increases, there is a consequent increase in the required preservative retention. The depth of penetration requirements may also increase.

The UCS was developed as a format change for the AWPA Commodity (or "C") Standards and is not intended to make substantive technical changes to those Standards. Inconsistencies in the UCS and the C Standards may exist due to the enormity of the task to construct the UCS. Therefore, in the event of conflicts between the UCS and the C Standards during the interim phase-in period, the C Standards will govern until such time as the C Standards may be deleted from a future edition of the AWPA Book of Standards. For user convenience, treated wood specification guidelines in this publication include both the appropriate Use Category Standard and a cross reference to its corresponding Commodity Standard.

To specify a treated wood commodity using the UCS, use the AWPA Book of Standards. The user should first find the appropriate Use Category for the expected service conditions in Section 2 "Service Conditions for Use Category Designations", and a definite application in Section 3 "Guide to Treated Wood End Uses". The user may then refer to the appropriate product or application in Section 7 "Commodity Specification." The Table of Contents for Commodity Specifications using AWPA's Use Category System is shown to the right.

COMMODITY SPECIFICATIONS
USE CATEGORY SYSTEM

Sawn Products	A
Permanent Wood Foundations (PWF)	A
Posts	B
Playground Material	B
Round Building Poles	B
Crossties	C
Utility Poles	D
Round Timber Piling	E
Wood Composites	F
Marine Applications (Salt Water)	G
Fire Retardants	H
Non-Pressure Applications	I

- Copper azole (CA)
- Sodium borate (SBX)

As of December 31, 2003, CCA was withdrawn for most residential consumer-use treated lumber applications. CCA treatment continues to be allowed for certain industrial, agricultural, foundation, and marine applications. More information on alternative preservatives and links to preservative manufacturers can be found on the Southern Pine Council website.

In consultation with the Environmental Protection Agency, manufacturers have made a transition to alternative wood preservatives for the residential and outdoor market. Leading preservative manufacturers including

- ArchWood Protection, Inc.
- Chemical Specialties, Inc.
- Osmose, Inc.

have amended their respective registrations for chromated copper arsenate (CCA), limiting the use of CCA to approved industrial and commercial applications.

Plywood recommended for the PWF system is all-veneer APA-rated plywood sheathing, Exposure 1-marked APA Series V-600, or Exterior-marked APA Series V-611, produced according to U.S. Product Standards PS1, PS2, or APA Standard PRP-108. The APA trademarks signify that the manufacturer is committed to APA's rigorous program of quality inspection and testing, and that panel quality is subject to verification through APA audit. Always insist that the plywood you use or specify for Permanent Wood Foundation bears the trademark "APA–The Engineered Wood Association."

CONSTRUCTING THE PERMANENT WOOD FOUNDATION

PWF construction is similar to wood-frame exterior wall construction, with some exceptions. Because PWF walls are used in below-grade applications, all lumber and plywood is pressure-treated with preservatives for decay and termite resistance. These have already been discussed. Other differences include the use of stainless steel nails, an offset footing plate, and framing anchors to connect foundation studs and floor joists to the top plates of foundation walls in high backfill conditions.

Like conventional wood-frame walls, the wood foundation is adaptable to virtually any design. It fits a variety of floor plans. This type of foundation can be used for both level and sloping sites.

Radon Gas

In certain localities where emission of radon gas from the soil or ground water is prevalent, a plastic pipe and tee can be installed through the basement floor for basement-type PWFs. For crawl space PWFs, a perforated plastic pipe can be installed on the ground inside the crawl space, beneath the vapor retarder. In both applications, the pipe is connected to a vent pipe and exhaust fan to depressurize the soil under the basement floor or crawl space vapor retarder, removing radon gas from the soil under and around the building. If a sump is used, the sump cover should be sealed and connected to the vent pipe and exhaust fan to remove radon gas from the sump pit.

ADVANTAGES OF THE PWF

Builders and home buyers across the country are choosing the Permanent Wood Foundation. Here are just a few of the features that are making PWFs increasingly popular:

Flexibility

Permanent Wood Foundations can be used in a variety of building types and sizes, including both single- and multistory houses, condominiums, and apartments, and for both site-built and manufactured houses. PWFs are suitable for crawl space, split-entry, or full-basement designs. Remodeling contractors have found the PWF ideal for room additions, especially where site access is limited.

Offices or other commercial and nonresidential buildings can also be built on a Permanent Wood Foundation. The PWF can be engineered for almost any large or complex building design or to satisfy special site constraints, and can be adapted for a variety of soil conditions, including low-bearing-capacity soils, expansive soil, or high water tables. The system can even be adapted for such uses as retaining walls and swimming pools.

Scheduling

The builder's or subcontractor's carpentry crews install the PWF, reducing the need for scheduling other trades. The PWF can be installed under nearly any weather conditions, even below freezing, so the building season is extended. On remote sites, high delivery costs and delays for concrete are eliminated. There's no need to wait for setting and stripping concrete forming, or allowing concrete to cure.

The PWF is easily installed by a small crew, often in less than a day or even in just a few hours. As soon as the foundation is framed and sheathed, construction of floors and walls can proceed. Shorter construction time means savings in interim construction financing—and greater productivity.

Comfortable Basement Several features of the PWF make a home attractive to buyers.

- First, there is comfort. Permanent Wood Foundation basements have all the livability of above-ground rooms. Wood construction lends a feeling of warmth. A damp feeling is usually associated with masonry basements. PWFs incorporate superior drainage features that prevent moisture problems typical of ordinary foundations. The result is a warm, dry, below-grade living space.

- Second, the wood-framed walls of the Permanent Wood Foundation make it easy to install thick, economical insulation. That means less heat loss through the foundation wall and greater long-term savings. The cost of installing insulation is less because the wood-framed walls are already in place. Consider this comparison: In a concrete foundation, ¾- or 1½-inch foam sheathing is typically installed for insulation. This gives an energy rating of R-3 to R-6. In the wood foundation, it's easy to fit 3½-inch-thick insulation between PWF studs, producing an R-11 to R-15 energy rating. If 5½-inch insulation is installed between 2 × 6 or 2 × 8 PWF studs, the energy rating increases to R-19 or R-21. The National Energy Policy Act mandates that the basement of a new home must be properly insulated. Several states have already adopted this code.

The economical answer to meeting state energy code requirements begins with a Permanent Wood Foundation. Research has found that to build an 8-inch basement wall with an insulation value of R-19, the concrete costs some 30% more than the PWF. Cement block can cost up to 62% more. Energy savings and an incomparably dry basement are possible. Comfortable living areas below grade are also possible.

Further Advantages of the PWF

Finishing An advantage of PWFs is the ease of finishing. The easily worked with wood studs are already in place. That means plumbing, wiring, and interior wall installation are simplified. Because it's so easy, many PWF home buyers elect to do the finishing themselves according to their own tastes and imagination—often at less cost.

More Space Permanent Wood Foundations can also mean added living space—wood foundation walls

need not be as thick as comparable concrete or masonry walls. Less space is needed for insulation, too, because it fits into the cavities of the wood-framed wall; thus, extra furring strips or wall studs aren't necessary. PWFs are easy to remodel or modify. Window or door openings can be cut out or whole rooms can be added. Additional structural engineering may be required for certain remodeling projects.

Radon Gas PWF systems, both new construction and retrofits, have definite advantages for radon gas resistance. The gravel layer beneath the basement floor serves as a collection system for soil gas. The gas is easily vented to the outside.

Soil Conditions

The type of soil and general grading conditions at the building site are factors in determining foundation construction details such as footing design, backfill, and drainage provisions.

Soils are classified by their composition and how they drain. Table 9-4 lists common soil types and their properties. Soil classifications for most areas are listed in the standard series of soil surveys published by the U.S. Department of Agriculture's Soil Conservation Service.

PWFs may be built in Group I, II, or III soils. In poorly drained Group III soils, granular fill under the slab for basement-type foundations must be at least 8 inches deep as opposed to the 4-inch minimum for Group I and II soils.

In such soil conditions, it may be more practical to build an above-grade crawl space foundation/floor system, especially for sites having a high water table, or where extreme amounts of rain often fall in short periods. Regardless of soil type, above-grade crawl space foundation/floor systems have the cost benefit of minimum excavation and backfill.

Group IV soils are generally unsatisfactory for wood foundations—that is, unless special measures are taken.

For building sites in regions where expansive clay soils in Groups II, III, or IV occur, a licensed soils engineer should be consulted to determine modifications required for foundation footings, drainage, soil moisture control, and backfill around the foundation. In such cases, special design considerations and construction details may be needed to avoid soil expansion or shrinkage that might otherwise affect foundation and floor performance.

For basement-type foundations, a sump draining to daylight or into a storm sewer or other stormwater drainage system is recommended for all soil groups.

Table 9-4 *Types of Soils and Related Design Properties*

Soil Group	Unified Soil Classification Symbol	Soil Description	Allowable Bearing in Pounds per Square Foot with Medium Compaction or Stiffness[1]	Drainage Characteristic[2]	Frost Heave Potential	Volume Change Potential Expansion[3]
Group I Excellent	GS	Well-graded gravels, gravel-sand mixtures, little or no fines.	8000	Good	Low	Low
	GP	Poorly graded gravels or gravel-sand mixtures, little or no fines.	8000	Good	Low	Low
	SW	Well-graded sands, gravelly sands, little or no fines.	6000	Good	Low	Low
	SP	Poorly graded sands or gravelly sands, little or no fines.	5000	Good	Low	Low
	GM	Silty gravels, gravel-sand-silt mixtures.	4000	Medium	Medium	Low
	SM	Silty sand, sand-silt mixtures.	4000	Medium	Medium	Low
Group II Fair to Good	GC	Clayey gravels, gravel-sand-clay mixtures.	4000	Medium	Medium	Low
	SC	Clayey sands, sand-clay mixture.	4000	Medium	Medium	Low
	ML	Inorganic silts and very fine sands, rock flour, silty or clayey fine sands or clayey silts with slight plasticity	2000	Medium	High	Low
	CL	Inorganic clays of low to medium plasticity, gravelly clays, sandy clays, silty clays, lean clays.	2000	Medium	Medium	Medum
Group III Poor	CH	Inorganic clays of high plasticity, fat clays.	2000	Poor	Medium	High
	MH	Inorganic silts, micaceous or diatomaceous fine sandy or silty soils, elastic silts.	2000	Poor	High	High
Group IV Unsatisfactory	OL	Organic silts and organic silty clays of low plasticity.	400	Poor	Medium	Medium
	OH	Organic clays of medium to high plasticity, organic silts	–0–	Unsatisfactory	Medium	High
	Pt	Peat and other highly organic soils.	–0–	Unsatisfactory	Medium	High

[1]Allowable bearing value may be increased 25 percent for very compact, coarse-grained gravelly or sandy soils or very stiff fine-grained clayey or silty soils. Allowable bearing value may be decreased 25 percent for loose, coarse-gained gravelly or sandy soils, or soft, fine-grained clayey or silty soils.

[2] The percolation rate for good drainage is over 4 inches per hour, medium drainage is 2 to 4 inches per hour, and poor is less than 2 inches per hour.

[3] For expansive soils, contact local soils engineer for verification of design assumptions.

[4] Dangerous expansion might occur if these soil types are dry but subject to future wetting.

In addition, for all types of foundations in all soil groups, the ground surface around the foundation should be graded to slope ½-inch per foot away from the structure. The backfill should be free of organic material, voids, or chunks of clay. It should be compacted and no more permeable than the surrounding soil.

Site Preparation

Site clearing and excavation methods for the Permanent Wood Foundation are the same as for conventional foundation systems. Organic materials including tree stumps or other vegetation should be removed and topsoil separated from excavated earth, which may be used later for backfiring or grading.

After clearing the site, it's wise to use a plot plan to locate foundation footings and trenches for plumbing, sewer, gas and electrical lines, and drainage trenches. Excavation for foundation footings, plumbing and other service and drainage trenches must be completed before soil treatment for termite protection, if required. Otherwise, retreatment is necessary.

After utility trenches are dug to the desired level, they can be lined with fine gravel or sand before pipes and conduit are set in place. Then the trenches are filled the rest of the way with gravel, coarse sand, or crushed rock. See Figs. 9-45, A and B.

Footings and Backfill

Granular materials are recommended footings under foundation walls, for fill under the basement slab or treated wood basement floor. They are also good as a portion of the backfill to provide an optimum drainage system to keep the underfloor area and foundation walls dry. The granular material may be crushed stone, gravel, or sand. It must be clean and free of silt, clay, and organic material. The size limitations are

- Maximum of ½ inch for crushed stone
- Maximum of ¾ inch for gravel
- Minimum of 1/16 inch for sand

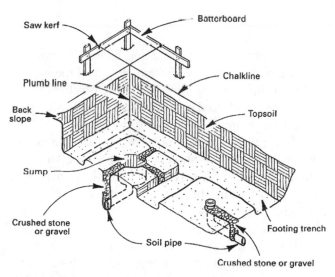

Fig. 9-45A *Typical excavation for footings and underfloor tile, pipes, conduit, and sumps.*

Continuous poured concrete may be used for the footings beneath foundation walls. If a concrete footing is used, it should be placed on gravel to maintain continuity of the drainage system. Otherwise, drains through the concrete footing must be provided.

All types of foundations need footings to be placed on undisturbed soil. The footing excavation should extend below the frost line. The footing trench depth and width, as with conventional systems, depend on the loads to be carried by the foundation. See Fig. 9-46.

Excavations must be wide and deep enough into undisturbed soil so that the footings will be centered under the foundation walls. Use of granular or concrete footings distributes vertical loads from the structure and foundation walls to the soil. Footings are required under perimeter and interior load-bearing walls.

Site Drainage

Site drainage is an important feature in keeping any type of foundation dry and trouble free. Drainage systems have been developed for wood foundations to keep crawl spaces and basements dry under virtually any condition. Granular footings and backfill are key elements in the PWF drainage system. They provide an unobstructed path for the water to flow away from the foundation, or into a sump for a basement house. This prevents a buildup of pressure against the foundation and helps avoid leaks. See Figs. 9-46 and 9-47.

Following are several methods of providing drainage around excavated foundations.

In permeable Group I soils, such as gravel, sand, silty gravels, and gravel-silt-sand mixtures with a percolation rate of more than 4 inches per hour, the gravel footing of the Permanent Wood Foundation provides

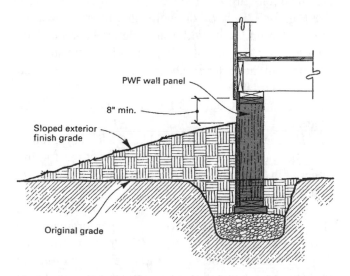

Fig. 9-45B *Above-grade crawl space foundation for high water table or soil drainage problems.*

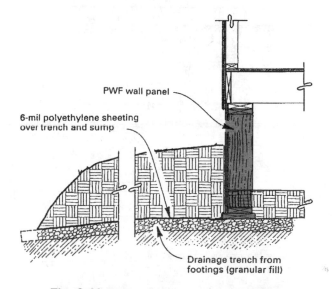

Fig. 9-46 *Footing drainage trench to daylight.*

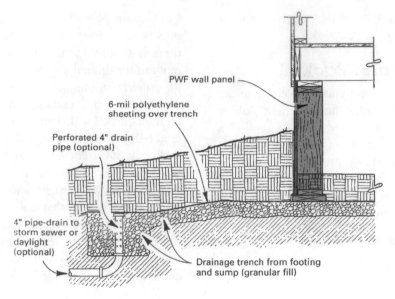

Fig. 9-47 *Footing drainage trench to sump or storm sewer.*

the drainage trench for sites with good surface water runoff. See Figs. 9-48 and 9-49.

For continuous concrete footings, place gravel under and around the outside of the footing, at least 12 inches deep. Water will collect in the gravel beneath the footing and then can be drained away from the foundation. Cover the gravel with 6-mil polyethylene sheeting or wrap gravel with water-permeable filter fabric to prevent soil from washing into the footing. See Fig. 9-50.

Group II soils such as gravel-sand-clay mixtures, clay, gravels or sands, inorganic sills, and fine sand

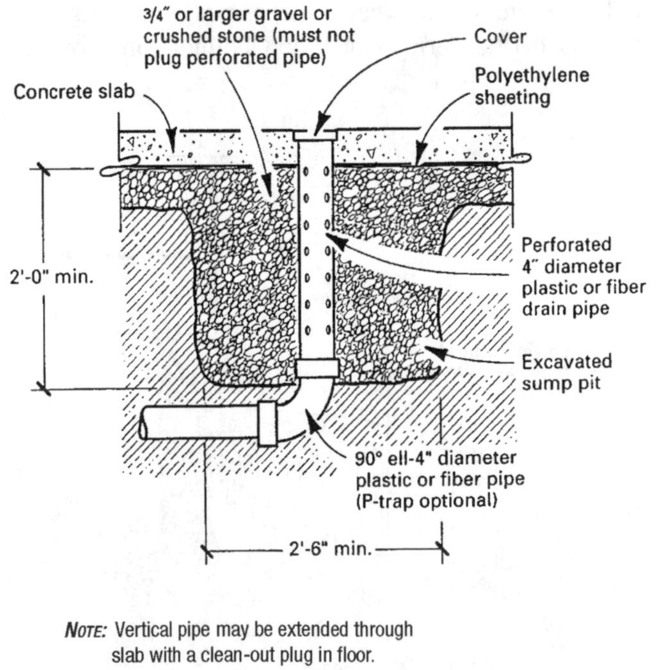

with medium-to-poor percolation characteristics, require a trench sloping away from the gravel foundation footing. If the site is sloping, it may be practical to dig a drainage trench to daylight, where the site slope intersects the drainage trench. See Fig. 9-46.

Place about 6 inches of gravel in the trench and cover with 6-mil polyethylene sheeting. An alternative is to place a drain tile or soil pipe into the trench from the gravel footing to the point where the trench emerges.

Foundation footings and basement sump, when used, should be drained to a storm sewer, drainage swale, or to daylight. On level sites where direct drainage to daylight or a storm sewer is not practical, it may be necessary to dig drywells or sump pits at several locations around the outside of the foundation. See Fig. 9-47. Rely on drywells only in areas of well-drained Group I soils with sand or gravel content. The bottom of the drywell should project into undisturbed, porous soil at a level above the highest seasonal groundwater table.

The sump pit, about 3 feet in diameter, should be at least 2 feet deeper than the base of the gravel footing. The top of the sump should be at a lower elevation than the footing. Connect the sump pit to the footing either with a gravel-filled trench or with a drainage tile or a soil pipe.

In medium-drained soils, a gravel-filled sump should be sufficient to provide proper drainage. In poorly drained soils, either provide drainage from the sump pit to a storm sewer or, in extreme drainage problem cases, install a sump pump inside prefabricated sump tile. See Fig. 9-47.

NOTE: Vertical pipe may be extended through slab with a clean-out plug in floor.

Fig. 9-48 *Sump for medium wet-drained soils.*

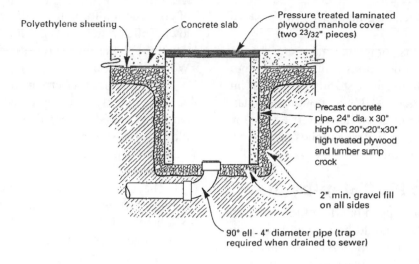

Polyethylene sheeting

Concrete slab

Pressure treated laminated
plywood manhole cover
(two 23/32" pieces)

Precast concrete
pipe, 24" dia. x 30"
high OR 20"x20"x30"
high treated plywood
and lumber sump
crock

2" min. gravel fill
on all sides

90° ell - 4" diameter pipe (trap
required when drained to sewer)

NOTE: Use of sump pump is required when sump
cannot be drained by gravity to daylight or sewer.

Fig. 9-49 *Sump for poorly drained soils.*

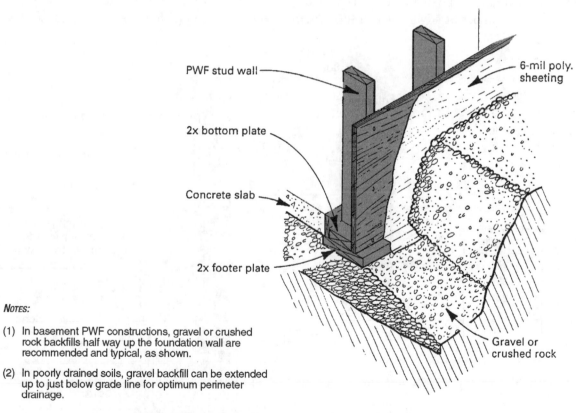

PWF stud wall

2x bottom plate

Concrete slab

2x footer plate

6-mil poly.
sheeting

Gravel or
crushed rock

NOTES:

(1) In basement PWF constructions, gravel or crushed
rock backfills half way up the foundation wall are
recommended and typical, as shown.

(2) In poorly drained soils, gravel backfill can be extended
up to just below grade line for optimum perimeter
drainage.

Fig. 9-50 *Basement PWF with perimeter gravel drainage.*

For crawl space foundations where the interior ground level is below outside finish grade, granular drainage trenches or drain pipes are recommended for draining footings or perimeter drains by gravity to daylight, storm sewers, or other approved stormwater drainage system.

On sites where proper drainage may be expensive or troublesome, consider using an above-grade foundation floor system and make sure the finish grade slopes away from the foundation.

In rainy climates, provide for drainage inside the foundation. This can be done by grading to a low spot on the ground inside such that the under-floor area will drain to it.

After the building is complete, make sure that the foundation and under-floor areas remain dry by providing for adequate drainage of stormwater. Most important is the use of gutters, down spouts and splash blocks, or drainpipe to direct water runoff away from the building. Also, slope adjacent porches or patios to drain away from the building.

Building the PWF Step-by-Step

There are a number of steps in building the Permanent Wood Foundation. They should be studied before undertaking the project, either to build it or to have it built. In fact, it would be best to access the entire plans in all their details from the website of the Southern Pine Council at *www.southernpine.com*. A complete 55-page manual is available at no charge for one copy.

The information shown here, in Figs. 9-51 through 9-65, give you enough detail to make a decision to go ahead with this alternative type of foundation. As you can see, there are a number of advantages of this type of construction for homes, offices, and stores. The following illustrations along with the captions will walk you through the entire process of building a Permanent Wood Foundation.

Installation of a Permanent Wood Foundation can be done quickly. On a prepared site, a PWF typically can be installed in less than one day. This makes it ideal for use where there is a problem with the weather. Experienced workers can install one in just a few hours.

The 17 steps shown in Figure 9-66A illustrate the rapidity with which the Permanent Wood Foundation can be completed. Figure 9-66B shows a completed basement. Note the placement of treated plywood sheathing.

Figures 9-67 and 9-68 show the advantage of the PWF when it comes to finishing the interior and exterior foundation walls to blend with the rest of the house.

Finishing the PWF House

It is easy to finish the interior and exterior foundation walls of a PWF structure. The walls can be finished to blend with the rest of the house. The exterior

h_i (in.)	Max. h_o (in.)
6	17
8	23
10	28
12	33
14	37

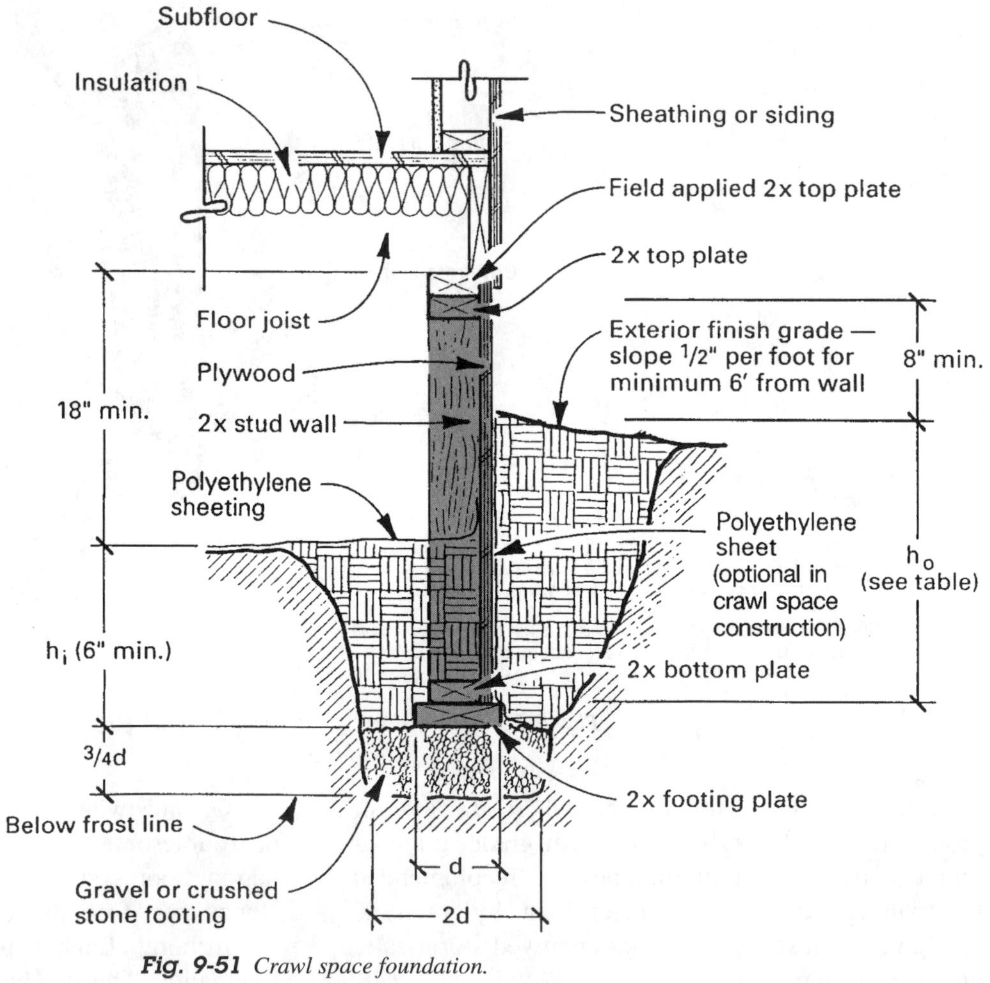

Fig. 9-51 *Crawl space foundation.*

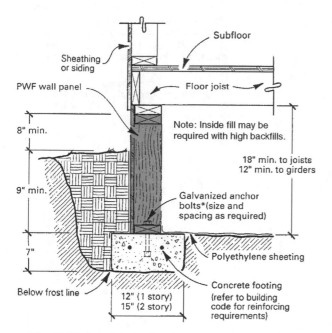

Fig. 9-52 *Crawl space foundation on concrete footing.*

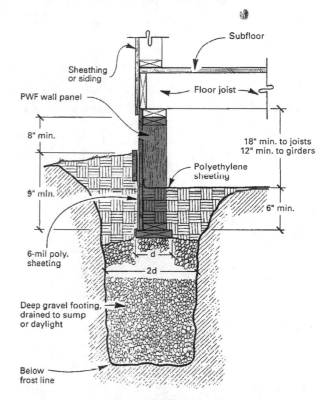

Fig. 9-53 *Deep gravel footing for PWF.*

wood foundation walls can be overlaid with APA-rated siding. The siding can be patterned southern pine exterior siding or other types of code-approved siding that blend with the architectural style of the structure. See Fig. 9-68. The wood siding should be at least 6 inches from the ground. Applying the siding is done, as with all types of housing, as shown in Fig. 9-67.

The exterior foundation walls can be stained, painted, or covered with a code-approved stucco finish. Film-forming, oil-based finishes are not recommended. They tend to crack and flake quickly in areas such as knotholes.

However, penetrating semitransparent oil-based stains generally perform well over CCA-treated plywood because they allow some of the wood to show through. The finished color may be affected by the color of the panel surface. Earth tones or a green stain will usually mask any discoloration from the chemical treatment.

The latex solid color stains or paint systems also show excellent performance over CCA pressure-treated plywood. Earth tones usually provide the best appearance. However, if pastels or white finishes are used, a stain-blocking all-acrylic latex primer, followed by a compatible all-acrylic latex topcoat, produces optimum results.

Plywood treated with ACA may contain blotchy deposits of residual surface salts. Thorough brushing to remove these excess surface salts prior to finishing is essential. Because of the blotchy appearance of the ACA treatment, only an all-acrylic latex paint system is recommended. It should be comprised of at least one coat of stain-blocking acrylic latex primer followed by a companion acrylic latex topcoat. Medium to darker colors are suggested.

The Southern Pine Council provided the illustrations for this last part of Chapter 9.

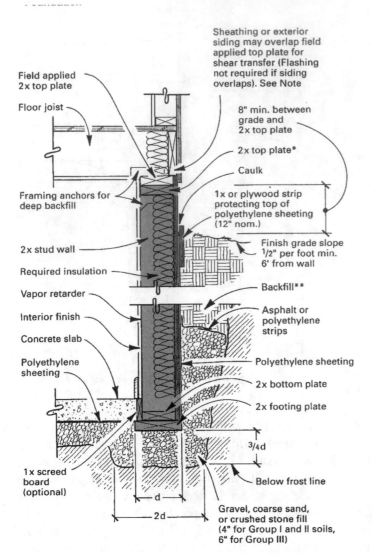

Field applied 2x top plate

Floor joist

Framing anchors for deep backfill

2x stud wall

Required insulation

Vapor retarder

Interior finish

Concrete slab

Polyethylene sheeting

1x screed board (optional)

Sheathing or exterior siding may overlap field applied top plate for shear transfer (Flashing not required if siding overlaps). See Note

8" min. between grade and 2x top plate

2x top plate*

Caulk

1x or plywood strip protecting top of polyethylene sheeting (12" nom.)

Finish grade slope 1/2" per foot min. 6' from wall

Backfill**

Asphalt or polyethylene strips

Polyethylene sheeting

2x bottom plate

2x footing plate

3/4 d

Below frost line

Gravel, coarse sand, or crushed stone fill (4" for Group I and II soils, 6" for Group III)

d

2d

* Not required to be treated if backfill is more than 8″ below bottom of plate. Typical for all details

** Backfill with crushed stone or gravel 12″ for Group I soils, and half the backfill height for Groups II and III soils.

NOTE: For daylight basement foundations, use double header joists (stagger end joints) or splice header joist for continuity on uphill and day-light sides of building.

Fig. 9-54 *Basement foundation.*

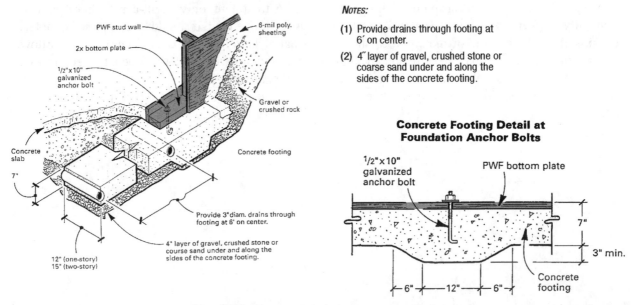

PWF stud wall

2x bottom plate

1/2"x10" galvanized anchor bolt

Concrete slab

7"

6-mil poly. sheeting

Gravel or crushed rock

Concrete footing

Provide 3" diam. drains through footing at 6' on center.

4" layer of gravel, crushed stone or course sand under and along the sides of the concrete footing.

12" (one-story)
15" (two-story)

NOTES:

(1) Provide drains through footing at 6′ on center.

(2) 4″ layer of gravel, crushed stone or coarse sand under and along the sides of the concrete footing.

Concrete Footing Detail at Foundation Anchor Bolts

1/2"x10" galvanized anchor bolt

PWF bottom plate

7"

3" min.

Concrete footing

6"   12"   6"

Fig. 9-55 *Basement foundation on concrete footing.*

Joist Perpendicular to Foundation Wall

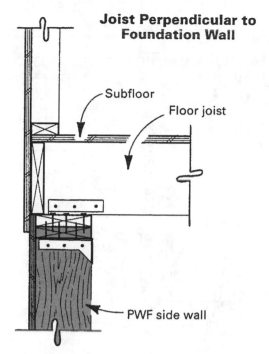

Subfloor

Floor joist

PWF side wall

Fig. 9-56 *Fastening foundation side walls to floor system.*

Joists Parallel to Foundation Wall

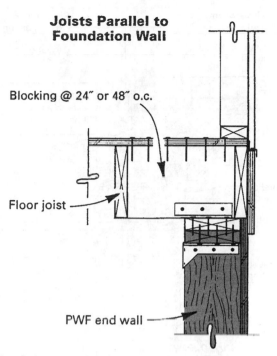

Blocking @ 24″ or 48″ o.c.

Floor joist

PWF end wall

Fig. 9-57 *Fastening foundation end walls to floor system.*

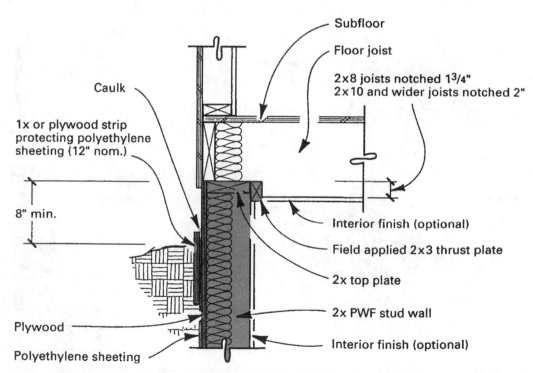

Caulk

1x or plywood strip protecting polyethylene sheeting (12" nom.)

8" min.

NOTE: Nailing to be as normally required for shallow or no fill. Lateral soil forces transferred through 2 x 3 thrust plate.

Plywood

Polyethylene sheeting

Subfloor

Floor joist

2x8 joists notched 1³/4"
2x10 and wider joists notched 2"

Interior finish (optional)

Field applied 2x3 thrust plate

2x top plate

2x PWF stud wall

Interior finish (optional)

Fig. 9-58 *Fastening foundation stud walls to floor system.*

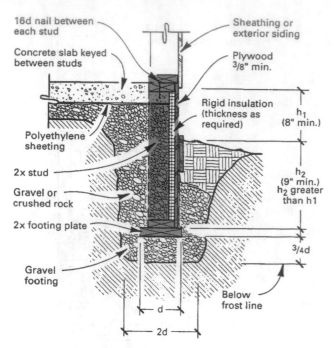

Fig. 9-59 *Daylight basement end walls.*

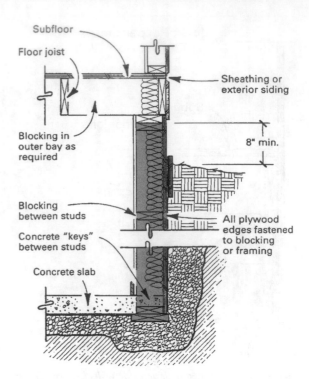

Fig. 9-60 *PWF stem wall, concrete slab on grade.*

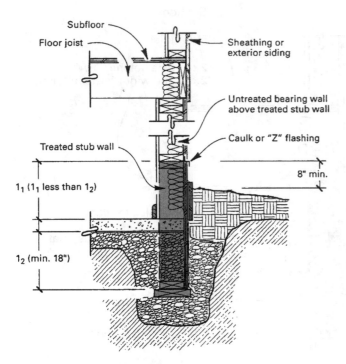

Fig. 9-61 *Basement foundation stub wall with low backfill.*

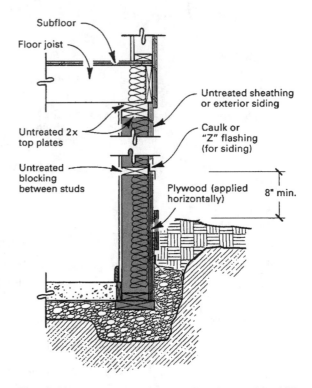

Fig. 9-62 *Basement foundation wall with partial backfill.*

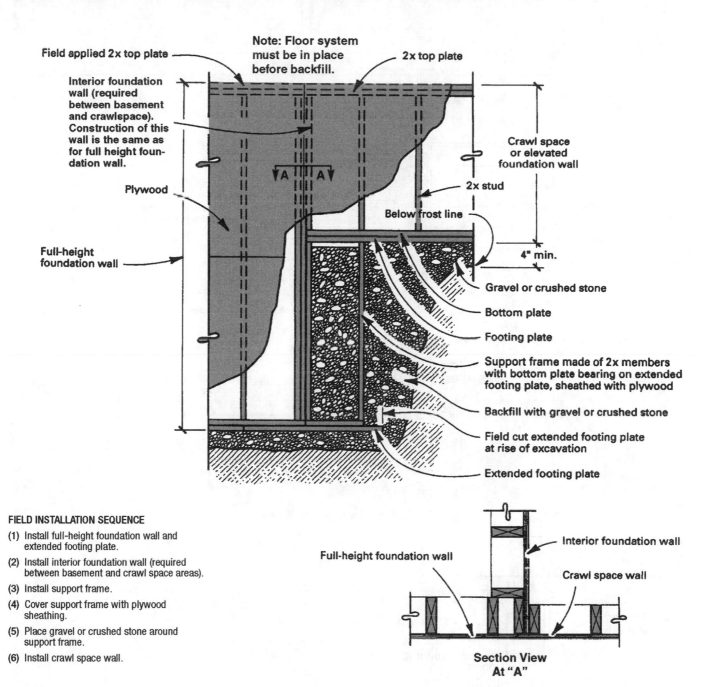

Field applied 2x top plate

Note: Floor system must be in place before backfill.

2x top plate

Interior foundation wall (required between basement and crawlspace). Construction of this wall is the same as for full height foundation wall.

Plywood

Full-height foundation wall

Crawl space or elevated foundation wall

2x stud

Below frost line

4" min.

Gravel or crushed stone

Bottom plate

Footing plate

Support frame made of 2x members with bottom plate bearing on extended footing plate, sheathed with plywood

Backfill with gravel or crushed stone

Field cut extended footing plate at rise of excavation

Extended footing plate

FIELD INSTALLATION SEQUENCE

(1) Install full-height foundation wall and extended footing plate.
(2) Install interior foundation wall (required between basement and crawl space areas).
(3) Install support frame.
(4) Cover support frame with plywood sheathing.
(5) Place gravel or crushed stone around support frame.
(6) Install crawl space wall.

Full-height foundation wall

Interior foundation wall

Crawl space wall

Section View At "A"

Fig. 9-63 Stepped-footing, crawl space, or basement foundation.

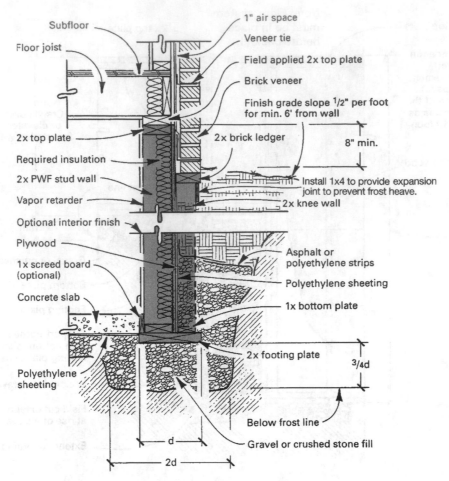

Fig. 9-64 *Knee-wall with brick veneer.*

Labels (top figure):
- Subfloor
- Floor joist
- 2x top plate
- Required insulation
- 2x PWF stud wall
- Vapor retarder
- Optional interior finish
- Plywood
- 1x screed board (optional)
- Concrete slab
- Polyethylene sheeting
- 1" air space
- Veneer tie
- Field applied 2x top plate
- Brick veneer
- Finish grade slope 1/2" per foot for min. 6' from wall
- 2x brick ledger
- 8" min.
- Install 1x4 to provide expansion joint to prevent frost heave.
- 2x knee wall
- Asphalt or polyethylene strips
- Polyethylene sheeting
- 1x bottom plate
- 2x footing plate
- Below frost line
- Gravel or crushed stone fill
- 3/4d
- d
- 2d

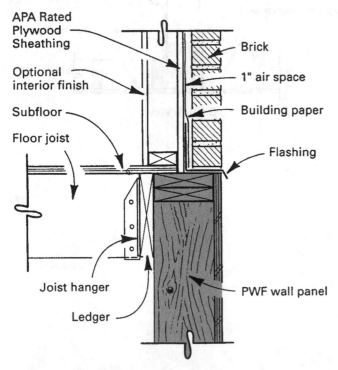

Fig. 9-65 *Alternate brick veneer detail.*

Labels (bottom figure):
- APA Rated Plywood Sheathing
- Optional interior finish
- Subfloor
- Floor joist
- Joist hanger
- Ledger
- Brick
- 1" air space
- Building paper
- Flashing
- PWF wall panel

Installation of a Permanent Wood Foundation is speedy. On a prepared site, a PWF can typically be installed in less than one day, but it's not unusual to see one installed in just hours. Here's the installation sequence:

1 The building site is prepared. Topsoil is removed and all excavation and trenching completed. Utility and drainage lines have been installed.

2 A minimum of 4″ of gravel, coarse sand, or crushed rock is laid as a base for the concrete slab or the wood floor to be installed later. Thickness of gravel under footings is relative to their width.

3 Gravel is leveled, extending several inches beyond where treated wood footing plates will rest. The gravel under the wood footing plates performs the same function as a conventional concrete spread footing, receiving and distributing loads from foundation walls.

4 Restaking the house is next, after drainage system and gravel footing installation.

5 Foundation sections can be built at the jobsite, or be prefabricated in panelized sections for accurate and rapid installation. Each section is composed of a footing plate, bottom plate, wood studs, plywood, and single top plate. Here, prefabricated sections are delivered ready for installation.

6 The first section goes up. 8′ x 8′ panels can be easily set without mechanical assistance. It is recommended that the first two panels installed be located at a corner, because a corner is self-bracing. Check the level of first sections

7 Caulking is applied between plywood edges of adjoining foundation sections. In full-basement construction, the plywood joints must be sealed full-length with caulking compound. Caulking is not required in crawl space construction.

8 Additional panels of the foundation are attached. Using these preframed sections, windows and door openings are already cut and framed, reducing onsite labor costs.

9 Bracing supports foundation panels while additional sections are installed.

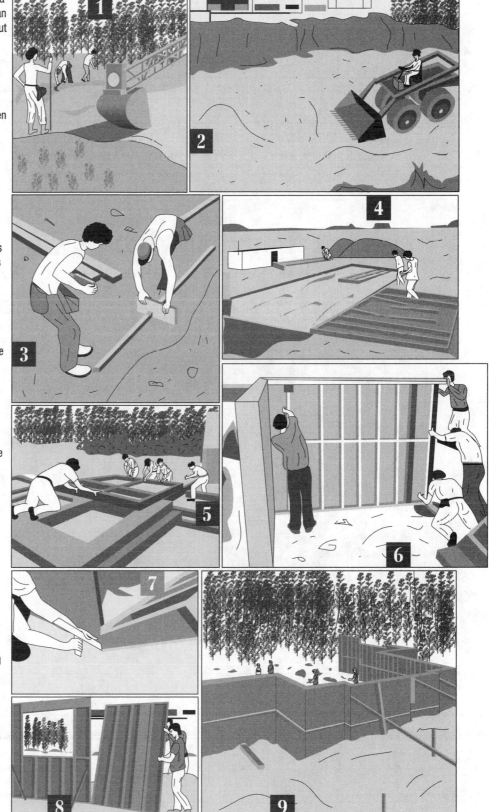

Fig. 9-66A *Illustrates how quickly the Permanent Wood Foundation can be completed.*

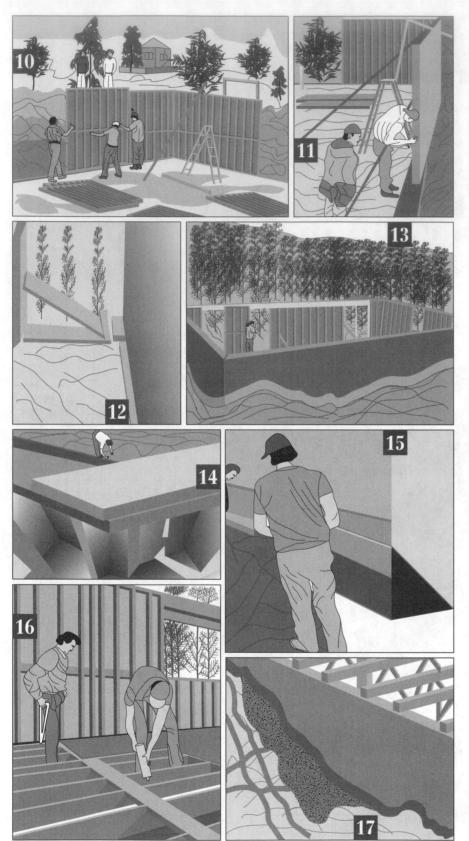

10 Remaining sections of foundation walls are erected. Extended footing plate automatically offsets wall section joints.

11 Panels are plumbed and aligned. When an accurate line has been established, sections can be shifted inward or outward as needed from the line of reference.

12 Stakes along the footing plate keep the panels from sliding until permanent attachment of all sections is completed.

13 The last section is installed. Once all the foundation wall sections are in place, the entire structure is rechecked to be level and square.

14 The second top plate (typically untreated lumber) is attached. Top plate joints are staggered so that they do not fall directly over joints between foundation wall sections. At corners, joints in the double top plates overlap as in conventional wall construction.

15 Prior to backfilling, 6-mil polyethylene sheeting is attached at the grade line and is draped over the portion of the foundation wall that will be below grade. The top edge is protected with a treated wood strip that is caulked. This strip is a guide for backfilling.

16 Installation of a treated wood floor will optimize the comfort of below-ground living areas.

17 Backfilling takes place after the basement floor is poured and cured, and the first floor framing and floor sheathing are installed. These steps give the PWF lateral restraint for backfill loads. To avoid excessive deflection, backfill in layers of 6″ to 8″ and tamp to compact. Avoid operating heavy equipment near walls during backfilling.

In Group I soils, the first 12″ or more of backfill is the same material as used for footings. For Group II and III soils, backfill with the same materials as footings, for half the height of the backfill. This portion of the fill is covered with strips of 30-pound asphalt paper or 6-mil polyethylene to permit water seepage, yet prevent infiltration of fine soils. Verify that no polyethylene sheeting is exposed below the grade strip.

Fig. 9-66A (*Continued*)

Fig. 9-66B *A complete basement. Note the placement of treated plywood sheathing.*

Fig. 9-68 *The PWF exterior foundation walls blend with the rest of the house.*

Fig. 9-67 *Application of siding.*

Appendices

A. Concrete Blocks and/or Concrete Masonry Units

B. Things to Know About Bricks

C. Welded Wire Fabric Reinforcement for Concrete

D. Specialty Plywood Panels

E. English to Metric Conversions

Appendix A. CONCRETE BLOCKS AND/OR CONCRETE MASONRY UNITS

Concrete blocks are hollow masonry units with a compressive strength of 600 to 1,500 psi. See Fig. A-1. This is a stretcher block that has two or three cores and nominal dimensions of 8 inches × 16 inches × 4 inches. Other units with 6-, 10-, and 12-inch widths are available:

- *Normal-weight* block is made from concrete weighing more than 125 pcf.
- *Medium-weight* block is made from concrete weighing from 105 to 125 pcf.
- *Lightweight* block is made from concrete weighing 105 pcf or less.

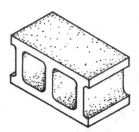

Fig. A-1 *Concrete block.*

Grades of Concrete Masonry Units

- *Grade N* is a load-bearing concrete masonry unit (CMU). It is suitable for use both above and below grade in walls exposed to moisture or weather. Grade N units have a compressive strength from 800 to 1,500 psi.
- *Grade S* is a load-bearing concrete masonry unit limited to use above grade, in exterior walls with weather-protective coatings, or in walls not exposed to moisture or weather. Grade S units have a compressive strength from 600 to 1,000 psi.

Types of Concrete Masonry Units

- *Type I* is a concrete masonry unit manufactured to a specific limit of moisture content in order to minimize the drying shrinkage that causes cracking.
- *Type II* is a concrete masonry unit not manufactured to any specific limit in moisture content.

Concrete Brick

Concrete brick is a solid, rectangular concrete masonry unit usually easily recognized since it is the size of a modular clay brick. It, too, is available in 12-inch

Fig. A-2 *Concrete brick.*

lengths. Concrete brick units have a compressive strength of 2,000 to 3,000 psi. See Fig. A-2.

Concrete Masonry Units

These are precast of a mixture of portland cement, fine aggregate, and water. They are molded into various shapes to satisfy specific construction conditions. See Fig. A-3A to A-3T. It illustrates the various shapes. Various types are available, with the local manufacturer deciding which to make for the area due to its demand. Concrete block is often incorrectly referred to as cement block.

Concrete block shapes and uses are as follows (see Fig. A-3):

A. Bull-nose blocks have one or more rounded exterior corners.

B. Corner blocks have a solid end face for use in constructing the end or corner of a wall.

C. Corner-return blocks are used as the corners of 6-, 10-, and 12-inch walls to maintain horizontal coursing with the appearance of full- and half-length units.

D. Double-corner blocks have solid faces at both ends. They are used in constructing a masonry pier.

E. Plaster blocks are used in constructing a plain or reinforced masonry pilaster.

F. Coping blocks are used in constructing the top or finishing course of a masonry wall.

G. Sash or jamb blocks have an end slot or rabbet to receive the jamb of a door or window frame.

H. Sill blocks have a wash to shed rainwater from a sill.

I. Cap blocks have a solid top for use as a bearing surface in the finishing course of a foundation wall.

J. Control-joint blocks are used in constructing a vertical control joint.

K. Sound-absorbing masonry units have a solid top and a slotted face shell. They sometimes use a fibrous filler, for increased sound absorption.

L. Bond-beam blocks have a depressed section in which reinforcing steel can be placed for embedment in grout.

M. Open-end blocks have one end open in which vertical steel reinforcement can be placed for embedment in grout.

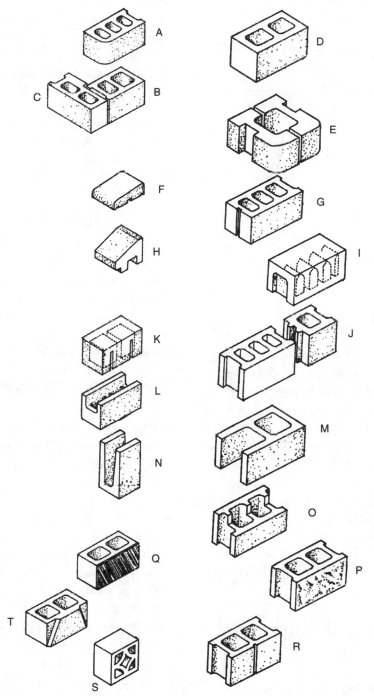

Fig. A-3 *Various shapes of concrete building blocks.*

N. Lintel blocks have a U-shaped section in which reinforcing steel can be placed for embedment in grout.

O. Header blocks have a portion of one face shell removed to receive headers in a bonded masonry wall.

P. Split-face blocks are split lengthwise by a machine after curing to produce a rough, fractured face texture.

Q. Faced blocks have a special ceramic, glazed, or polished face.

R. Scored blocks have a face shell with a pattern of beveled recesses.

S. Screen blocks are used in tropical architecture. They have a decorative pattern of transverse openings for admitting air and excluding sunlight.

T. Shadow blocks have a shell with a pattern of beveled recesses.

Concrete is normally specified according to the compressive strength it will develop within 28 days

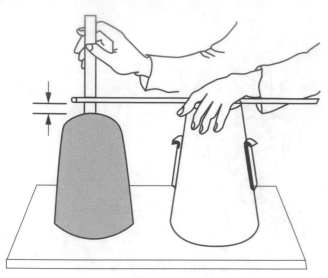

Fig. A-4 *Slump test.*

after placement (7 days for high-early-strength concrete). See Fig. A-4.

- Slump test is a method for determining the consistency and workability of freshly mixed concrete by measuring the slump of a test specimen, expressed as the vertical settling, in inches, of a specimen after it has been placed in a slump cone, tamped in a prescribed manner, and the cone is lifted.

- Compression test for determining the compressive strength of a concrete batch uses a hydraulic press to measure the maximum load a test cylinder 6 inches (150 millimeters) and 12 inches (305 millimeters) high can support in axial compression before fracturing. See Tables A-1 and Appendix E for conversion factors.

Water-Cement Ratio

The *water-cement ratio* is the ratio of mixing water to cement in a unit volume of a concrete mix, expressed by weight as a decimal fraction or as gallons of water per sack of cement. The water-cement ratio controls the strength, durability, and watertightness of hardened concrete. According to Abram's law, formulated by D. A. Abrams in 1919 from experiments at the Lewis Institute in Chicago, the compressive strength of concrete is inversely proportional to the ratio of water to cement. If too much water is used, the concrete mix will be weak and porous after curing. If too little water is used, the mix will be dense but difficult to place and work. For most applications, the water-cement ratio should range from 0.45 to 0.60. See Table A-2.

Steel Reinforcement

Because concrete is relatively weak in tension, reinforcement consisting of steel bars, strands, or wires is required to absorb tensile, shearing, and sometimes the compressive stresses in a concrete member or structure. Steel reinforcement is also required to tie vertical and horizontal elements, reinforce the edges around openings, minimize shrinkage cracking, and control thermal expansion and contraction. All reinforcement should be designed by a qualified structural engineer. See Fig. A-5.

- Reinforcing bars are steel sections hot-rolled with ribs or other deformations for better mechanical bonding to concrete. The bar number refers to its diameter in eighths of an inch; for example, a no. 5 bar is ⅝ inch (16 millimeters) in diameter.

- Welded wire fabric consists of a grid of steel wires or bars welded together at all points of intersection. The fabric is typically used to provide temperature reinforcement for slabs, but the heavier gages can also be used to reinforce concrete walls. The fabric is designated by the size of the grid in inches followed by a number indicating the wire gage or cross-sectional area.

- Reinforcing steel must be protected by the surrounding concrete against corrosion and fire. Minimum requirements for cover and spacing are specified by the American Concrete Institute (ACI) *Building Code Requirements for Reinforced Concrete* according to the concrete's exposure and the size of the coarse aggregate and steel used. See Table A-3.

Appendix B. THINGS TO KNOW ABOUT BRICKS
Bricks

The term *masonry* refers to building with units of various natural or manufactured products. These can be brick, stone, or concrete block. They are usually used

Table A-1 *Metric Conversion Factors*

Factor	Multiples	Prefixes	Symbols
One thousand million	10^9	giga	G
One million	10^6	mega	M
One thousand	10^3	kilo	k
One hundred	10^2	hecto	h
Ten	10	deka	da
One-tenth	10^{-1}	deci	d
One-hundredth	10^{-2}	centi	c
One-thousandth	10^{-3}	milli	m
One-millionth	10^{-6}	micro	μ

Table A-2 *Concrete Strength vs. Water-Cement Ratio*

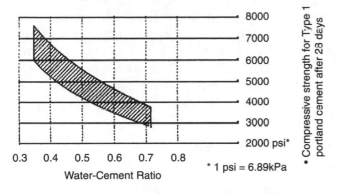

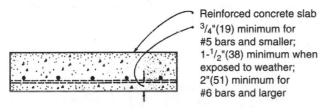

Reinforced concrete slab
$^3/_4$"(19) minimum for
#5 bars and smaller;
1-$^1/_2$"(38) minimum when
exposed to weather;
2"(51) minimum for
#6 bars and larger

Fig. A-5 *Reinforced concrete slab.*

Table A-3 *ASTM Standard Reinforcing Bars*

| | Nominal Dimensions | | |
| | | Cross-Sectional | |
Bar Size	Diameter, in. (mm)	Area, in.² (mm²)	Weight, plf* (N/m)
No. 3	0.375 (10)	0.11 (71)	0.38 (5.5)
No. 4	0.50 (13)	0.20 (129)	0.67 (9.7)
No. 5	0.625 (16)	0.31 (200)	1.04 (15.2)
No. 6	0.75 (19)	0.44 (284)	1.50 (21.9)
No. 7	0.875 (22)	0.60 (387)	2.04 (29.8)
No. 8	1.00 (25)	0.79 (510)	2.67 (39.0)
No. 9	1.125 (29)	1.00 (645)	3.40 (49.6)
No. 10	1.25 (32)	1.27 (819)	4.30 (62.8)

* Pounds per linear foot.

with mortar as a bonding agent. Uniform sizes and proportional relationships of units of masonry distinguish them from most of the other building materials. Because the masonry unit is structurally most effective in compression, the masonry unit should be laid up in such a way that the entire masonry mass acts as an entity. That is where bricklaying becomes important in the building of sturdy structures.

Common brick is also called *building brick*. It is made for general building purposes and is not specially treated for color and/or texture. *Face brick* is made of special clays for facing a wall. It is often treated to produce the desired color and surface texture. See Fig. B-1.

Types of Brick

The brick type designates the permissible variation in size, color, chippage, and distortion that is allowed in the face brick. Three types of brick are FBX, FBS, and FBA.

- FBX is face brick that is suitable for use where
  - Minimum variation in size is permitted
  - Narrow color range is required
  - A high degree of mechanical perfection is required.
- FBS is face brick that is suitable for use where
  - A wider range of color is permitted
  - A greater variation in size is permitted than for the FBX type.
- FBA is face brick that is suitable for use where effects are desired that result from nonuniformity in size, color, and texture of the individual bricks.

Efflorescence is a white, powdery deposit that forms on exposed masonry and concrete surfaces. It is caused by the leaching and crystallization of soluble salts from within the material. Reducing moisture absorption is the best assurance against efflorescence.

Brick is a masonry unit made of clay. It is formed into a rectangular prism while plastic and is hardened by firing in a kiln or drying in the sun. Five methods are used for making brick:

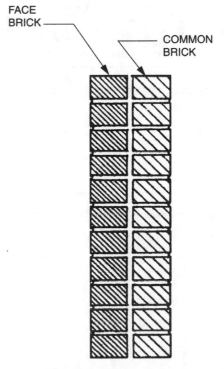

Fig. B-1 *Use of face brick.*

1. *Soft mud process.* Brick is formed by molding relatively wet clay. The clay has a moisture content of 20% to 30%.

2. *Sand struck.* Brick is formed in the soft-mud process with a mold lined with sand. This is done to prevent sticking. It produces a matted-texture surface.

3. *Water struck.* Brick is formed in the soft-mud process with a mold lubricated with water. This prevents sticking. It produces a smooth, dense surface.

4. *Stiff-mud process.* Brick is formed by extruding stiff, but plastic clay. The clay has a moisture content of 12% to 15%. It is pushed through a die and extruded. It is cut to length with wires before it is fired in the kiln.

5. *Dry-press process.* Brick is formed by molding relatively dry clay. The clay has a moisture content of 5% to 7%. It is under high pressure. This results in a sharp-edged brick with a smooth surface.

Actual dimensions of bricks vary due to shrinkage during the manufacturing process. The nominal dimensions are given in Table B-1.

Table B-1 Brick Dimensions

Brick Unit	Nominal Dimensions (Thickness × Height × Length) in.	mm	Modular Coursing (C) in.	mm
Modular	4 × 2⅔ × 8	100 × 68 × 205	3C* = 8	205
Norman	4 × 2⅔ × 12	100 × 68 × 305	3C = 8	205
Engineer	4 × 3⅕ × 8	100 × 81 × 205	5C =16	405
Norwegian	4 × 3⅕ × 12	100 × 81 × 305	5C =16	405
Roman	4 × 2 × 12	100 × 51 × 305	2C = 4	100
Utility	4 × 4 × 12	100 × 100 × 305	1C = 4	100

*Courses.

Grades of Brick

The grade of a brick is designated according to its durability. Exposure to weathering is the primary concern here. In regard to the grading of brick, the United States is divided into three weathering regions: severe, moderate, and negligible. The type relates to the annual winter rainfall and the annual number of freezing-cycle days. Brick is graded for use in each region according to

- Compressive strength
- Maximum water absorption
- Maximum saturation coefficient

The brick is labeled SW, MW, or SW. These refer to the localities suggested for their use. For instance:

- SW is brick suitable for exposure to *severe weathering.* This means it is in contact with the ground or used on surfaces likely to be permeated with water in subfreezing temperatures. It has a minimum compressive strength of 2,500 psi.

- MW is brick suitable for exposure to *moderate weathering.* This means it is used above grade on surfaces unlikely to be permeated with water in subfreezing temperatures. It has a minimum compressive strength of 2,200 psi.

- NW is brick suitable for exposure to negligible weathering. This means it is used as a backup or in interior masonry. Its minimum compressive strength is 1,250 psi.

Allowable compressive stresses in masonry walls are much less than the values shown here. That is so because the quality of the masonry units, mortar, and workmanship may vary.

Appendix C.   Welded Wire Fabric Reinforcement for Concrete

Table C-1   *Welded Wire Fabric Reinforcement for Concrete*

Type of Construction	Recommended Style	Remarks
Barbecue Foundation Slab	6 × 6-8/8 to 4 × 4-6/6	Use heavier style fabric for heavy, massive fireplaces or barbecue pits.
Basement Floors	6 × 6-10/10, 6 × 6-8/8 or 6 × 6-6/6	For small areas (15-foot maximum side dimension) use 6 × 6-10/10. As a rule of thumb, the larger the area or the poorer the sub-soil, the heavier the gage.
Driveways	6 × 6-6/6	Continuous reinforcement between 25- to 30-foot contraction joints.
Foundation Slabs (Residential only)	6 × 6-10/10	Use heavier gauge over poorly drained sub-soil, or when maximum dimension is greater than 15 feet.
Garage Floors	6 × 6-6/6	Position at midpoint of 5- or 6-inch-thick slab.
Patios and Terraces	6 × 6-10/10	Use 6 × 6-8/8 if sub-soil is poorly drained.
Porch Floor		
a. 6-inch-thick slab up to 6-foot span	6 × 6-6/6	Position 1 inch from bottom
b. 6-inch-thick slab up to 8-foot span	4 × 4-4/4	form to resist tensile stresses.
Sidewalks	6 × 6-10/10 6 × 6-8/8	Use heavier gauge over poorly drained sub-soil. Construct 25- to 30-foot slabs as for driveways.
Steps (Free span)	6 × 6-6/6	Use heavier style if more than five risers. Position fabric 1 inch from bottom form.
Steps (On ground)	6 × 6-8/8	Use 6 × 6-6/6 for unstable sub-soil.

Appendix D.   Specialty Plywood Panels

Table D-1   *Specialty Plywood Panels.*

Panel Type and Trademark	Description, Bond Classification, and Thickness
APA Decorative **APA** *THE ENGINEERED WOOD ASSOCIATION* DECORATIVE GROUP 1 EXPOSURE 1 ——— 000 ——— PS 1-95	Rough sawn, brushed, grooved, or striated faces. For paneling, interior accent walls, built-ins, counter facing, exhibit displays. Can also be made by some manufacturers in Exterior for exterior siding, gable ends, fences, and other exterior applications. Use recommendations for Exterior panels vary with the particular product. Check with the manufacturer. Bond Classifications: Interior, Exposure 1. Common Thicknesses: $5/16$, $3/8$, $1/2$, $5/8$.
APA High Density Overlay (HDO) HDO • A-A • EXT-APA • 000 • PS1-95	Has a hard, semi-opaque resin-fiber overlay on both faces. Abrasion-resistant. For concrete forms, cabinets, countertops, signs and tanks. Also available with skid-resistant screen-grid surface. Bond Classification: Exterior. Common Thicknesses: $3/8$, $1/2$, $5/8$, $3/4$.
APA Medium Density Overlay (MDO) **APA** *THE ENGINEERED WOOD ASSOCIATION* M.D. OVERLAY GROUP 1 EXTERIOR ——— 000 ——— PS 1-95	Smooth, opaque, resin-fiber overlay on one or both faces. Ideal base for paint, both indoors and outdoors. For exterior siding, paneling, shelving, exhibit displays, cabinets, signs. Bond Classification: Exterior. Common Thicknesses: $11/32$, $3/8$, $15/32$, $1/2$, $19/32$, $5/8$, $23/32$, $3/4$.
APA Marine MARINE • A-A • EXT-APA • 000 • PS1-95	Ideal for boat hulls. Made only with Douglas fir or western larch. Subject to special limitations on core gaps and face repairs. Also available with HDO or MDO faces. Bond Classification: Exterior. Common Thicknesses: $1/4$, $3/8$, $1/2$, $5/8$, $3/4$.
APA Plyform Class **APA** *THE ENGINEERED WOOD ASSOCIATION* PLYFORM **B-B** CLASS 1 EXTERIOR ——— 000 ——— PS 1-95	Concrete form grades with high reuse factor. Sanded both faces and mill-oiled unless otherwise specified. Special restrictions on species. Also available in HDO for very smooth concrete finish, and with special overlays. Bond Classification: Exterior. Common Thicknesses: $19/32$, $5/8$, $23/32$, $3/4$.
APA Plyron PLYRON • EXPOSURE 1• APA • 000	Hardboard face on both sides. Faces tempered, untempered, smooth, or screened, For countertops, shelving, cabinet doors, and flooring. Bond Classifications: Interior, Exposure 1. Common Thicknesses: $1/2$, $5/8$, $3/4$.

Courtesy of American Plywood Association.

Appendix E.   English to Metric Conversions

Table E-1   *English to Metric Conversions*

Measurement	Imperial Unit	Metric Unit	Metric Abbreviation	Conversion Factor
Length	mile	kilometer	km	1 mi = 1.609 km
	yard	meter	m	1 yd = 0.9144 m = 914.4 mm
	foot	meter	m	1 ft = 0.3408 m = 304.8 mm
		millimeter	mm	1 ft = 304.8 mm
	inch	millimeter	mm	1 in. = 25.4 mm
Area	square mile	square kilometer	km^2	1 mi^2 = 2.590 km^2
		hectare	ha	1 mi^2 = 259.0 ha (1 ha = 10,000 m^2)
	acre	hectare	ha	1 acre = 0.4047 ha
		square meter	m^2	1 acre = 4046.9 m^2
	square yard	square meter	m^2	1 yd^2 = 0.8361 m^2
	square foot	square meter	m^2	1 ft^2 = 0.0929 m^2
		square centimeter	cm^2	1 ft^2 = 929.03 cm^2
	square inch	square centimeter	cm^2	1 in.^{2} = 6.452 cm^2
Volume	cubic yard	cubic meter	m^3	1 yd^3 = 0.7646 m^3
	cubic foot	cubic meter	m^3	1 ft^3 = 0.02832 m^3
		liter	l	1 ft^3 = 28.32 l (1000 l = 1 m^3)
		cubic decimeter	dm^3	1 ft^3 = 28.32 dm^3 (1 l = 1 dm^3)
	cubic inch	cubic millimeter	mm^3	1 in.^{3} = 16,390 mm^3
		cubic centimeter	cm^3	1 in.^{3} = 16.39 cm^3
		milliliter	ml	1 in.^{3} = 16.39 ml
		liter	l	1 in.^{3} = 0.01639 l
Mass	ton	kilogram	kg	1 ton = 1,016.05 kg
	kip (1,000 lb)	metric ton (1,000 kg)	kg	1 kip = 453.59 kg
	pound	kilogram	kg	1 lb = 0.4536 kg
	ounce	gram	g	1 oz = 28.35 g
Per length	pound/lf	kilogram/meter	kg/m	1 plf = 1.488 kg/m
Per area	pound/sf	kilogram/meter^2	kg/m^2	1psf = 4.882 kg/m^2
Mass density	pound/cu ft	kilogram/meter^3	kg/m^3	1 pcf = 16,018 kg/m^3
Capacity	quart	liter	l	1 qt = 1.137 l
	pint	liter	l	1 pt = 0.568 l
	fluid ounce	cubic centimeter	cm^3	1 fl oz = 28.413 cm^3
Force	pound	newton	N	1 lb = 4.488 N
				1 N = 1 kg · m/s^2
Per length	pound/lf	newton/meter	N/m	1 plf = 14.594 N/m
Pressure	pound/sf	pascal	Pa	1 psf = 47.88 Pa
				1 Pa = 1 N/m^2
	pound/sq in.	kilopascal	kPa	1 psi = 6.894 kPa
Moment	foot-pound	newton-meter	N · m	1 ft · lb = 1.356 N · m
Mass	pound-feet	kilogram-meter	kg · m	1 lb · ft = 0.138 kg · m
Inertia	pound-feet^2	kilogram-meter^2	kg · m^2	1 lb · ft^2 = 0.042 kg · m^3
Velocity	miles/hour	kilometer/hour	km/h	1 mph = 1.609 km/h
	feet/minute	meter/minute	m/min	1 fpm = 0.3408 m/min
	feet/second	meter/second	m/s	1 fps = 0.3408 m/s
Volume rate of flow	cu ft/minute	liter/second	l/s	1 ft^3/min = 0.4791 l/s
	cu ft/second	meter^3/second	m^3/s	1 ft^3/s = 0.02832 m^3/s
	cu in./second	milliliter/second	ml/s	1 in.^{3}/s = 16.39 ml/s
Temperature	degree Fahrenheit	degree Celsius	°C	$t\,°C = \frac{5}{9}\,(t\,°F - 32)$
	degree Fahrenheit	degree Celsius	°C	1°F = 0.5556°C
Heat	British thermal unit (Btu)	joule	J	1 Btu = 1,055 J
		kilojoule	kJ	1 Btu = 1.055 kJ
Flow	Btu/hour	watt	W	1 Btu/h = 0.2931 W
Conductance	Btu · in/(ft^2 · h · °F)	watt/(meter^2 · °C)	W/(m^2 · °C)	1 Btu/(ft^2 · h · °F) = 5.678 W/(m^2 · °C)
Resistance	ft^2 · h · °F/Btu	meter^2 · K/W	m^2 · °C/W	1 ft^2 · h · °F/Btu = 0.176 m^2 · °C/W
Refrigeration	ton	watt	W	1 ton = 3,519 W
Power	horsepower	watt	W	1 hp = 745.7 W
		kilowatt	kW	1 hp = 0.7457 kW
Light	candela	candela	cd	Basic SI unit of luminous intensity
Lux	lumen	lumen	lm	1 lm = cd steradian
Illuminance	footcandle	lux	lx	1 fc = 10.76 lx
	lumen/sf	lux	lx	1 lm/ft^2 = 10.76 lx
Luminance	foot-lambert	candela/meter^2	cd/m^2	1 ft · L = 3.426 cd/m^2

Glossary

access Access refers to the freedom to move to and around a building, or the ease with which a person can obtain admission to a building site.

acoustical This refers to the ability of tiles on a ceiling to absorb or deaden sound.

aggregate Aggregate refers to sand, gravel, or both in reference to concrete mix.

air-drying Method of removing excess moisture from lumber using natural circulation of air.

alligatoring This occurs when paint cracks and resembles the skin of an alligator.

anchor bolt An anchor bolt is a steel pin that has a threaded end with a nut and an end with a 90° angle in it. The angled end is pushed into the wet concrete and becomes part of the foundation for anchoring the flooring or sill plates.

annular ring There are two colored rings that indicate the growth of a tree. The colors indicate the growth of springwood and summerwood.

apron A piece of window trim that is located beneath the window sill. Also used to designate the front of a building such as the concrete apron in front of a garage.

arbor An axle on which a cutting tool is mounted. It is a common term used in reference to the mounting of a circular saw blade.

asphalt shingle This is a composition-type shingle used on a roof. It is made of a saturated felt paper with ground-up pieces of stone embedded and held in place by asphaltum.

asphalt shingles These are shingles made of asphalt or tar-impregnated paper with a mineral material embedded; they are very fire resistant.

auxiliary locks Auxiliary locks are placed on exterior doors to prevent burglaries.

awl An awl is a tool used to mark wood with a scratch mark. It can be used to produce pilot holes for screws.

awning picture window This type of window has a bottom panel that swings outward; a crank operates the moving window. As the window swings outward, it has a tendency to create an awning effect.

backsplash A backsplash is the vertical part of a countertop that runs along the wall to prevent splashes from marring the wall.

backsaw This saw is easily recognized since it has a very heavy steel top edge. It has a fine-tooth configuration.

balloon frame This type of framing is used on two-story buildings. Wall studs rest on the sill. The joists and studs are nailed together, and the end joists are nailed to the wall studs.

balustrade A complete handrail assembly. This includes the rails, the balusters, subrails, and fillets.

baluster The baluster is that part of the staircase which supports the handrail or bannister.

bannister The bannister is that part of the staircase which fits on top of the balusters.

baseboard Molding covering the joint between a finished wall and the floor.

base shoe A molding added at the bottom of a baseboard. It is used to cover the edge of finish flooring or carpeting.

batten A batten is the narrow piece of wood used to cover a joint.

batter boards These are boards used to frame in the corners of a proposed building while the layout and excavating work takes place.

batts Batts are thick pieces of Fiberglas that can be inserted into a wall between the studs to provide insulation.

bay window Bay windows stick out from the main part of the house. They add to the architectural qualities of a house and are used mostly for decoration.

beam A horizontal framing member. It may be made of steel or wood. Usually the term is used to refer to a wooden beam that is at least 5 inches thick and at least 2 inches wider than it is thick.

bearing partition An interior divider or wall that supports the weight of the structure above it.

bearing wall A bearing wall has weight-bearing properties associated with holding up a building's roof or second floor.

benchmark A point from which other measurements are made.

bevel A bevel is a tool that can be adjusted to any angle. It helps make cuts at the number of degrees that is desired and is a good device for transferring angles from one place to another.

bifold A bifold is a folding door used to cover a closet. It has two panels that hinge in the middle and fold to allow entrance.

blistering Blistering refers to the condition that paint presents when air or moisture is trapped underneath and makes bubbles that break into flaky particles and ragged edges.

blocking Corners and wall intersections are made the same as outside walls. The size and amount of blocking can be reduced. The purpose of blocking is to provide nail surfaces at the corners. These are needed at inside and outside nail surfaces. They are a base for nailing wall covering.

blockout A form for pouring concrete is blocked out by a frame or other insertion to allow for an opening once the concrete has cured.

blocks This refers to a type of flooring made of wood. Wide pieces of boards are fastened to the floor, usually in squares and by adhesives.

board and batten A finished wall surface consisting of vertical board with gaps between them. The battens or small strips of wood cover the gaps.

board foot A unit of lumber measure equaling 144 cubic inches. The base unit (B.F.) is 1 inch thick and 12 inches square, or $1 \times 12 \times 12 = 144$ cubic inches.

bonding This is another word for gluing together wood or plastics and wood.

bottom or heel cut This refers to the cutout of the rafter end which rests against the plate. The bottom cut is also called the foot or seat cut.

bow A term used to indicate an upward warp along the length of a piece of lumber that is laid.

bow window A window unit that projects from an exterior wall. It has a number of windowpanes that form a curve.

brace A brace is an inclined piece of lumber applied to a wall or to roof rafters to add strength.

brace scale A brace scale is a table that is found along the center of the back of the tongue and gives the exact lengths of common braces.

bridging Bridging is used to keep joists from twisting or bending.

buck A buck is the same as a blockout.

builder's level This is a tripod-mounted device that uses optical sighting to make sure that a straight line is sighted and that the reference point is level.

building codes Building codes are rules and regulations which are formulated in a code by a local housing authority or governing body.

building paper Also called tar paper, roofing paper, and a number of other terms. It has a black coating of asphalt for use in weatherproofing.

building permits Most incorporated cities or towns have a series of permits that must be obtained for building. This allows for inspections of the work and for placing the house on the tax rolls.

built-ins This is a term used to describe the cabinets and other small appliances that are built into the kitchen, bathroom, or family room by a carpenter. They may be custom cabinets or may be made on the site.

bundle This term refers to the packaging of shingles. A bundle of shingles is usually a handy method for shipment.

butt A term that can be used a couple of ways: A *butt hinge* is one where the two parts meet edge-to-edge, allowing movement of the two parts when held together by a pin; to *butt* means to meet edge to edge, such as in a joining of wooden edges.

cantilever Overhangs are called *cantilevers*; they are used for special effects on porches, decks, or balconies.

carpenter's square This steel tool can be used to check for right angles, to lay out rafters and studs, and to perform any number of measuring jobs.

carpet strips These are wooden or metal strips with nails or pins sticking out. They are nailed around the perimeter of a room, and the carpet is pulled tight and fastened to the exposed nails.

carpet tape This is a tape used to seam carpet where it fits together.

carriage A notched stair frame is called a carriage.

casement This is a type of window hinged to swing outward.

casing A door casing refers to the trim that goes on around the edge of a door opening and also to the trim around a window.

cathedral ceiling A cathedral ceiling is not flat and parallel with the flooring; it is open and follows the shape of the roof. The open ceiling usually precludes an attic.

caulk Caulk is any type of material used to seal walls, windows, and doors to keep out the weather. Caulk is usually made of putty or some type of plastic material, and it is flexible and applied with a caulking gun.

cellulose fiber Insulation material made from cellulose fiber. Cellulose is present in wood and paper, for example.

cement Cement is a fine-powdered limestone that is heated and mixed with other minerals to serve as a binder in concrete mixes.

ceramic tile Ceramic tile is made of clay and fired to a high temperature; it usually has a glaze on its surface. Small pieces are used to make floors and wall coverings in bathrooms and kitchens.

certificate of occupancy This certificate is issued when local inspectors have found a house worthy of human habitation. It allows a contractor to sell a house. It is granted when the building code has been complied with and certain inspections have been made.

chair A chair is a support bracket for steel reinforcing rods that holds the rods in place until the concrete has been poured around them.

chalk line A chalk line is used to guide a roofer. It is snapped, causing the string to make a chalk mark on the roof so that the roofers can follow it with their shingles.

chipboard Chipboard is used as an underlayment. It is constructed of wood chips held together with different types of resins.

chisel A wood chisel is used to cut away wood for making joints. It is sharpened on one end, and the other is hit with the palm of the hand or with a hammer to cut away wood for door hinge installation or to fit a joint tightly.

claw hammer This is the common hammer used by carpenters to drive nails. The claws are used to extract nails that bend or fail to go where they are wanted.

cleat Any strip of material attached to the surface of another material to strengthen, support, or secure a third material.

clerestory A short exterior wall between two sections of roof that slope in different directions. The term is also used to describe a window that is placed in this type of wall.

closed-cut For a closed-cut valley, the first course of shingles is laid along the eaves of one roof area up to and over the valley. It is extended along the adjoining roof section. The distance is at least 12 inches. The same procedure is followed for the next courses of shingles.

cold chisel This chisel is made with an edge that can cut metal. It has a one-piece configuration, with a head to be hit by a hammer and a cutting edge to be placed against the metal to be cut.

common rafter A common rafter is a member that extends diagonally from the plate to the ridge.

concrete Concrete is a mixture of sand, gravel, and cement in water.

condensation The process by which moisture in the air becomes water or ice on a surface (such as a window) whose temperature is colder than the air's temperature.

contact cement Contact cement is the type of glue used in applying countertop finishes. Both sides of the materials are coated with the cement, and the cement is allowed to dry. The two surfaces are then placed in contact, and the glue holds immediately.

contractor A contractor is a person who contracts with a firm, a bank, or another person to do a job for a certain fee and under certain conditions.

contractor's key This is a key designed to allow the contractor access to a house while it is under construction. The lock is changed to fit a pregrooved key when the house is turned over to the owner.

convection Transfer of heat through the movement of a liquid or gas.

coped joint This type of joint is made with a coping saw. It is especially useful for corners that are not square.

coping saw This saw is designed to cut small thicknesses of wood at any curve or angle desired. The blade is placed in a frame, with the teeth pointing toward the handle.

corner beads These are metal strips that prevent damage to drywall corners.

cornice The cornice is the area under the roof overhang. It is usually enclosed or boxed in.

course This refers to alternate layers of shingles in roofing.

cradle brace A cradle brace is designed to hold sheetrock or drywall while it is being nailed to the ceiling joists. The cradle brace is shaped like a T.

crawl space A crawl space is the area under a floor that is not fully excavated. It is only excavated sufficiently to allow one to crawl under it to get at the electrical or plumbing devices.

cricket This is another term for the *saddle*.

cripple jack A cripple jack is a jack rafter with a cut that fits in between a hip and a valley rafter.

cripple rafter A cripple rafter is not as long as the regular rafter used to span a given area.

cripple stud This is a short stud that fills out the position where the stud would have been located if a window, door, or some other opening had not been there.

crisscross wire support This refers to chicken wire that is used to hold insulation in place under the flooring of a house.

crosscut saw This is a handsaw used to cut wood across the grain. It has a wooden handle and a flexible steel blade.

cup To warp across the grain.

curtain wall Inside walls are often called curtain walls. They do not carry loads from roof or floors above them.

dado A rectangular groove cut into a board across the grain.

damper A damper is an opening and closing device that will close off the fireplace area from the outside by shutting off the flue. It can also be used to control draft.

deadbolt lock A deadbolt lock will respond only to the owner who knows how to operate it. It is designed to keep burglars out.

deck A deck is the part of a roof that covers the rafters.

decorative beams Decorative beams are cut to length from wood or plastic and mounted to the tastes of the owner. They do not support the ceiling.

diagonals Diagonals are lines used to cut across from adjacent corners to check for squareness in the layout of a basement or footings.

dividers Dividers have two points and resemble a compass. They are used to mark off specific measurements or transfer them from a square or a measuring device to the wood to be cut.

dormer Dormers are protrusions that stick out from a roof. They may be added to allow light into an upstairs room.

double-hung windows Double-hung windows have two sections, one of which slides past the other. They slide up and down in a prearranged slot.

double plate This usually refers to the practice of using two pieces of dimensional lumber for support over the top section or wall section.

double trimmer Double joists used on the sides of openings are called double trimmers. Double trimmers are placed without regard to regular joist spacings for openings in the floor for stairs or chimneys.

downspouts These are pipes connected to the gutter to conduct rainwater to the ground or sewer.

drain tile A drain tile is usually made of plastic. It generally is 4 inches in diameter, with a number of small holes to allow water to drain into it. It is laid along the foundation footing to drain the seepage into a sump or storm sewer.

drawer guide The drawers in cabinets have guides to make sure that the drawer glides into its closed position easily without wobbling.

drop siding Drop siding has a special groove, or edge, cut into it. The edge lets each board fit into the next board. This makes the boards fit together and resist moisture and weather.

drywall This is another name for panels made of gypsum.

ductwork Ductwork is a system of pipes used to pass heated air along to all parts of a house. The same ductwork can be used to distribute cold air for summer air conditioning.

dutch hip This is a modification of the hip roof.

eaves Eaves are the overhang of a roof projecting over the walls.

eaves trough A gutter.

elevation Elevation refers to the location of a point in reference to the point established with the builder's level or transit. Elevation indicates how high a point is. It may also refer to the front elevation or front view of a building or the rear elevation or what the building looks like from the rear. Side elevations refer to a side view.

energy Energy refers to the oil, gas, or electricity used to heat or cool a house.

essex board measure This is a table on the back of the body; it gives the contents of any size lumber. The table is located on the steel square used by carpenters.

excavation Excavate means to remove. In this case, excavation refers to the removal of dirt to make room for footings, a foundation, or the basement of a building.

expansion joint This is usually a piece of soft material that is about 1 inch thick and 4 inches wide and is placed between sections of concrete to allow for expansion when the flat surface is heated by the sun.

exposure Exposure refers to the part of a shingle or roof left to the weather.

extension form An extension form is built inside the concrete outer form. It forms a stepped appearance so that water will not drain into a building but drain outward from the slab or foundation slab.

faced insulation This insulation usually has a coating to create a moisture barrier.

faced stapling Faced stapling refers to the strip along the outer edges of the insulation that is stapled to the outside or 2-inch sides of the 2-×-4 studs.

facing Facing strips give cabinets a finished look. They cover the edges where the units meet and where the cabinets meet the ceiling or woodwork.

factory edge This is the straight edge of linoleum made at the factory. It provides a reference line for the installer.

factory-produced housing This refers to housing that is made totally in a factory. Complete units are usually trucked to a place where a basement or slab is ready.

false bottom This is a system of 1-×-6-inch false-bottom or box beams that provide the beauty of beams without the expense. The false beams are made of wood or plastic materials and glued or nailed in place. They do not support any weight.

fascia Fascia refers to a flat board covering the ends of rafters on the cornice or eaves. The eave troughs are usually mounted to the fascia board.

FHA The Federal Housing Administration.

Fiberglas Fiberglas is insulation material made from spun resin or glass. It conducts little heat and creates a large dead air space between layers of fibers. It helps conserve energy.

firebrick This is a special type of brick that is not damaged by fire. It is used to line the firebox in a fireplace.

fire stops Fire stops are short pieces nailed between joists and studs.

flaking This refers to paint that falls off a wall or ceiling in flakes.

flashing Flashing is metal used to cover chimneys or other things projecting through the roofing. It keeps the weather out.

floating The edges of drywall sheets are staggered, or floated. This gives more bracing to the wall since the whole wall does not meet at any one joint.

floating Floating refers to concrete work; it lets the smaller pieces of concrete mix float to the top. Floating is usually done with a tool moved over the concrete.

floorboards This refers to floor decking. Floorboards may be composed of boards, or may be a sheet of plywood used as a subfloor.

flue The flue is the passage through a chimney.

flush This term means to be even with.

folding rule This is a device that folds into a 3-×–6-inch rectangle. It has the foot broken into 12 inches. Each inch is broken into 16 points. Snap joints hinge the rule every six inches. It will spread to as much as 6 feet, or 72 inches.

footings Footings are the lowest part of a building. They are designed to support the weight of the building and distribute it to the earth or rock formation on which it rests.

form A form is a structure made of metal or wood used as a mold for concrete.

Formica Formica is a laminated plastic covering made for countertops.

foundation The foundation is the base on which a house or building rests. It may consist of the footings and walls.

framing Roof framing is composed of rafters, ridge board, collar beams, and cripple studs.

framing square This tool allows a carpenter to make square cuts in dimensional lumber. It can be used to lay out rafters and roof framing.

French doors This usually refers to two or more groups of doors arranged to open outward onto a patio or veranda. The doors are usually composed of many small glass panes.

frostline This is the depth to which the ground freezes in the winter.

furring strips These are strips of wood attached to concrete or stone. They form a nail base for wood or paneling.

gable This is the simplest kind of roof. Two large surfaces come together at a common edge, forming an inverted V.

galvanized iron This material is usually found on roofs as flashing. It is sheet metal coated with zinc.

gambrel roof This is a barn-shaped roof.

girder A girder is a support for the joists at one end. It is usually placed halfway between the outside walls and runs the length of the building.

grade The grade is the variation of levels of ground or the established ground-line limit on a building.

grid system This is a system of metal strips that support a drop ceiling.

grout Grout is a white plaster-like material placed into the cracks between ceramic tiles.

gusset A gusset is a triangular or rectangular piece of wood or metal that is usually fastened to the joint of a truss to strengthen it. It is used primarily in making roof trusses.

gutter This is a metal or wooden trough set below the eaves to catch and conduct water from rain and melting snow to a downspout.

gypsum Gypsum is a chalk used to make wallboard. It is made into a paste, inserted between two layers of paper, and allowed to dry. This produces a plastered wall with certain fire-resisting characteristics.

handsaw A handsaw is any saw used to cut wood and operated by manual labor rather than electricity.

hang a door This term refers to the fact that a door has to be mounted on hinges and aligned with the door frame.

hangers These are metal supports that hold joists or purlins in place.

hardboard A type of fiberboard pressed into thin sheets. Usually made of wood chips or waste material from trees after the lumbering process has been completed.

hardware In this case hardware refers to the metal parts of a door. Such things as hinges, locksets, and screws are hardware.

hardwood The wood that comes from a tree that sheds its leaves. This doesn't necessarily mean the wood itself is hard. A poplar has soft wood, but it is classified as a hardwood tree. An oak has hard wood and is also classified as a hardwood tree.

header A header is a board that fits across the ends of joists.

head lap This refers to the distance between the top and the bottom shingle and the bottom edge of the shingle covering it.

hearth A hearth is the part of a fireplace that is in front of the wood rack.

hex strips This refers to strips of shingles that are six-sided.

hip rafters A hip rafter is a member that extends diagonally from the corner of the plate to the ridge.

hip roof A hip roof has four sides, all sloping toward the center of the building.

hollow-core doors Most interior doors are hollow and have paper or plastic supports for the large surface area between the top and bottom edge and the two faces.

honeycomb Air bubbles in concrete cause a honeycomb effect and weaken the concrete.

insert stapling This refers to stapling insulation inside the 2-×-4 stud. The facing of the insulation has a strip left over. It can be stapled inside the studs or over the studs.

insulation Insulation is any material that offers resistance to the conduction of heat through its composite materials.

Plastic foam and Fiberglas are the two most commonly used types of insulation in homes today.

insulation batts These are thick precut lengths of insulation designed to fit between studs.

interlocking Interlocking refers to a type of shingle that overlaps and interlocks with its edges. It is used in high winds.

in the white This term is used to designate cabinets that are assembled but unfinished.

jack rafter Any rafter that does not extend from the plate to the ridge is called a jack rafter.

jamb A jamb is the part that surrounds a door window frame. It is usually made of two vertical pieces and a horizontal piece over the top.

joist Large dimensional pieces of lumber used to support the flooring platform of a house or building are called joists.

joist hangers These are metal brackets that hold up the joist. They are nailed to the girder, and the joist fits into the bracket.

joist header If the joist does not cover the full width of the sill, space is left for the joist header. The header is nailed to the ends of the joists and rests on the sill plate. It is perpendicular to the joists.

kerf The cut made by a saw blade.

key A key is a depression made in a footing so that the foundation or wall can be poured into the footing, preventing the wall or foundation from moving during changes in temperature or settling of the building.

kicker A kicker is a piece of material installed at the top or side of a drawer to prevent it from falling out of a cabinet when it is opened.

kiln dried Special ovens are made to dry wood before it is used in construction.

king-post truss This is the type of roof truss used to support low pitch roofs.

ladder jack Ladder jacks hang from a platform on a ladder. They are most suitable for repair jobs and for light work where only one carpenter is on the job.

landing A landing is the part of a stairway that is a shaped platform.

lap This refers to lap siding. Lap siding fits on the wall at an angle. A small part of the siding is overlapped on the preceding piece of siding.

laths Laths are small strips of wood or metal designed to hold plastic on the wall until it hardens for a smooth finish.

ledger A ledger is a strip of lumber nailed along an edge or bottom of a location. It helps support or keep from slipping the girders on which the joists rest.

left-hand door A left-hand door has its hinges mounted on the left when viewed from the outside.

level A level is a tool using bubbles in a glass tube to indicate the level of a wall, stud, or floor. Keeping windows, doors, and frames square and level makes a difference in their fit and operation.

level-transit This is an optical device that is a combination of a level and a means for checking vertical and horizontal angles.

load Load refers to the weight of a building.

load conditions These are the conditions under which a roof must perform. The roof has to support so much wind load and snow. Load conditions vary according to locale.

lockset The lockset refers to the doorknob and associated locking parts inserted in a door.

mansard This type of roof is popular in France and is used in the United States also. The second story of the house is covered with the same shingles used on the roof.

manufactured housing This term is used in reference to houses that are totally or partially made within a factory and then trucked to a building site.

mastic Mastic is an adhesive used to hold tiles in place. The term also refers to adhesives used to glue many types of materials in the building process.

military specifications These are specifications that the military writes for the products it buys from the manufacturers. In this case, the term refers to the specifications for a glue used in making trusses and plywood.

miter box The miter box has a hacksaw mounted in it. It is adjustable for cutting at angles such as 45 and 90°. Some units can be adjusted by a level to any angle.

modular homes These houses are made in modules or small units which are nailed or bolted together once they arrive at the foundation or slab on which they will rest.

moisture barrier A moisture barrier is some type of material used to keep moisture from entering the living space of a building. Moisture barrier, vapor seal, and membrane mean the same thing. It is laid so that it covers the whole subsurface area over sand or gravel.

moisture control Excess moisture in a well insulated house may pose problems. A house must be allowed to breathe and change the air occasionally, which in turn helps remove excess moisture. Proper ventilation is needed to control moisture in an insulated house. Elimination of moisture is another method but it requires the reduction of cooking vapors and shower vapors, for example.

moldings Moldings are trim mounted around windows, floors, and doors as well as closets. Moldings are usually made of wood with designs or particular shapes.

monolithic slab Mono means "one." This refers to a one-piece slab for a building floor and foundation all in one piece.

nail creep This is a term used in conjunction with drywall, where the nails pop because of wood shrinkage. The nailheads usually show through the panel.

nailers These are powered hammers that have the ability to drive nails. They may be operated by compressed air or by electricity.

nailhead stains These occur whenever the iron in the nailhead rusts and shows through the paint.

nail set Finish nails are driven below the surface of the wood by a nail set. The nail set is placed on the head of a nail, and the large end of the nail set is struck with a hammer.

newels The end posts of a stairway are called newels.

octagon scale This "eight square" scale is found on the center of the face of the tongue of a steel square. It is used when timber is cut with eight sides.

open This refers to the type of roofing that allows a joint between a dormer and the main roof. It is an open valley type of roofing. The valley where two roofs intersect is left open and covered with flashing and roofing sealer.

orbital sander This power sander will vibrate, but in an orbit. Thus causes the sandpaper to do its job better than it would if used in only one direction. An orbital sander can be used to finish off windows, doors, counters, cabinets, and floors.

panel This refers to a small section of a door that takes on definite shape, or to the panel in a window made of glass.

panel door This is a type of door used for the inside of a house. It has panels inserted in the frame to give it strength and design.

parquet Parquet is a type of flooring made from small strips arranged in patterns. It must be laminated to a base.

partition A partition is a divider wall or section that separates a building into rooms.

peeling This is a term used in regard to paint that will not stay on a building. The paint peels and falls off or leaves ragged edges.

penny (d) This is the unit of measure of the nails used by carpenters.

perimeter The perimeter is the outside edges of a plot of land or building. It represents the sum of all the individual sides.

perimeter insulation Perimeter insulation is placed around the outside edges of a slab.

pile A pile is a steel or wooden pole driven into the ground sufficiently to support the weight of a wall and building.

pillar A pillar is a pole or reinforced wall section used to support the floor and consequently the building. It is usually located in the basement, with a footing of its own to spread its load over a wider area than the pole would normally occupy.

pitch The pitch of a roof is the slant or slope from the ridge to the plate.

pivot This refers to a point where the bifold door is anchored and allowed to move so that the larger portion of the folded sections can move.

plane Planes are designed to remove small shavings of wood along a surface. One hand holds the knob in front, and the other hand holds the handle in the back of the plane. A plane is used to shave off door edges to make them fit properly.

planks This refers to a type of flooring usually made of tongue-and-groove lumber and nailed to the subflooring or directly to the floor joists.

plaster This refers to plaster of Paris mixed with water and applied to a lath to cover a wall and allow for a finished appearance that will take a painted finish.

plaster grounds A carpenter applies small strips of wood around windows, doors, and walls to hold plaster. These grounds may also be made of metal.

plastic laminates These are materials usually employed to make countertops. Formica is an example of a plastic laminate.

plate The plate is a roof member which has the rafters fastened to it at their lower ends.

platform frame This refers to the flooring surface placed over the joists; it serves as support for further floor finishing.

plenum A plenum is a large chamber.

plumb bob This is a very useful tool for checking plumb, or the upright level of a board, stud, or framing member. It is also used to locate points that should be directly under a given location. It hangs free on a string, and its point indicates a specific location for a wall, a light fixture, or the plumb of a wall.

plumb cut This refers to the cut of the rafter end which rests against the ridge board or against the opposite rafter.

post and beam Posts are used to support beams, which support the roof decking. Regular rafters are not used. This technique is used in barns and houses to achieve a cathedral-ceiling effect.

prehung This refers to doors or windows that are already mounted in a frame and are ready for installation as a complete unit.

primer This refers to the first coat of paint or glue when more than one coat will be applied.

pulls The handle or the part of the door on a cabinet or the handle on a drawer that allows it to be pulled or opened is called a pull.

purlin Secondary beams used in post-and-beam construction are called purlins.

rabbet A groove cut in or near the edge of a piece of lumber to fit the edge of another piece.

radial-arm saw This type of power saw has a motor and blade that moves out over the table which is fixed. The wood is placed on the table and the blade is pulled through the wood.

rafter scales This refers to a steel square with the rafter measurements stamped on it. The scales are on the face of the body.

rail The vertical facing strip on a cabinet is the stile. The horizontal facing strip is the rail.

rake On a gabled roof, a rake is the inclined edge of the surface to be covered.

random spacing This refers to spacing that has no regular pattern.

rebar A rebar is a reinforcement steel rod in a concrete footing.

reinforcement mesh Reinforcement mesh is made of 10-gage wires spaced about 4 to 6 inches apart. It is used to reinforce basements or slabs in houses. The mesh is placed so that it becomes a part of the concrete slab or floor.

remodeling This refers to changing the looks and function of a house.

residential building A residential building is designed for people to live in.

resilient flooring This type of flooring is made of plastics rather than wood products. It includes such things as linoleum and asphalt tile.

ridge board This is a horizontal member that connects the upper ends of the rafters on one side to the rafters on the opposite side.

right-hand door A right-hand door has the opening or hinges mounted on the right when viewed from the outside.

right-hand draw This means that the curtain rod can be operated to open and close the drapes from the right-hand side as one faces it.

rise In roofing, rise is the vertical distance between the top of the double plate and the center of the ridge board. In stairs, it is the vertical distance from the top of a stair tread to the top of the next tread.

riser The vertical part at the edge of a stair is called a riser.

roof brackets These brackets can be clamped onto a ladder used for roofing.

roof cement A number of preparations are used to make sure that a roof does not leak. Roof cement also can hold down shingle tabs and rolls of felt paper when it is used as a roof covering.

roofing This term is used to designate anything that is applied to a roof to cover it.

rough line A rough line is drawn on the ground to indicate the approximate location of footing.

rough opening This is a large opening made in a wall frame or roof frame to allow the insertion of a door or window or whatever is to be mounted in the open space. The space is shimmed to fit the object being installed.

router A router will cut out a groove or cut an edge. It is usually powered and has a number of different shaped tips that will carve its shape into a piece of wood. It can be used to take the edges off countertops.

run The run of a roof is the shortest horizontal distance measured from a plumb line through the center of the ridge to the outer edge of the plate.

R values This refers to the unit that measures the effectiveness of insulation. It indicates the relative value of the insulation for the job. The higher the number, the better the insulation qualities of the materials.

saber saw The saber saw has a blade that can be used to cut circles in wood. It can cut around any circle or curve. The blade is inserted in a hole drilled previously and the saw will follow a curved or straight line to remove the block of wood needed to allow a particular job to be completed.

saddle A saddle is the inverted V-shaped piece of roof inserted between the vertical side of a chimney and the roof.

saturated felt Other names for this material are tar paper and builder's felt. It is roll roofing paper and can be used as a moisture barrier and waterproofing material on roofs and under siding.

scabs Scabs are boards used to join the ends of a girder.

scaffold A scaffold is a platform erected by carpenters to stand on while they work on a higher level. Scaffolds are supported by tubing or 2 × 4s. Another name for scaffolding is *staging*.

screed A screed may be a board or pipe supported by metal pins. The screed is leveled with the tops of the concrete forms. It is removed after the section of concrete is leveled.

scribing Scribing means marking.

sealant A sealant is any type of material that will seal a crack. This usually refers to caulking when carpenters use the term.

sealer coat The sealer coat ensures that a stain is covered and the wood is sealed against moisture.

shakes This is a term used for shingles made of handsplit wood, in most cases western cedar.

sheathing This is a term used for the outside layer of wood applied to studs to close up a house or wall. It is also used to cover the rafters and make a base for the roofing. It is usually made of plywood today. In some cases, sheathing is still used to indicate the 1 × 6-inch wooden boards used for siding undercoating.

shed In terms of roofs, this is the flat sloping roof used on some storage sheds. It is the simplest type of roof.

sheetrock This is another name for panels made of gypsum.

shim To shim means to add some type of material that will cause a window or door to be level. Usually wood shingles are wedge-shaped and serve this purpose.

shingles This refers to material used to cover the outside of a roof and take the ravages of weather. Shingles may be made of metal, wood, or composition materials.

shingle stringers These are nailing boards that can have cedar shingles attached to them. They are spaced to support the length of the shingle that will be exposed to weather.

shiplap An L-shaped edge, cut into boards and some sheet materials to form an overlapping joint with adjacent pieces of the same material.

side lap The side lap is the distance between adjacent shingles that overlap.

siding This is a term used to indicate that the studs have been covered with sheathing and the last covering is being placed on it. Siding may be made of many different materials—wood, metal, or plastic.

sill This is a piece of wood that is anchored to the foundation.

sinker nail This is a special nail for laying subflooring. The head is sloped toward the shank but is flat on top.

size Size is a special coating used for walls before the wallpaper is applied. It seals the wall and allows the wallpaper paste to attach itself to the wall and paper without adding undue moisture to the wall.

skew-back saw This saw is designed to cut wood. It is hand-operated and has a serrated steel blade that is smooth on the non-cutting edge of the saw. It is 22 to 26 inches long and can have 5½ to 10 teeth per inch.

skilled worker A skilled worker is a person who can do a job well each time it is done, or the person who has the ability to do the job a little better each time. Skilled means the person has been at it for some time, usually 4 to 5 years at the least.

sliding door This is usually a large door made of glass, with one section sliding past the other to create a passageway. A sliding door may be made of wood or glass and can disappear or slide into a wall. Closets sometimes have doors that slide past one another to create an opening.

sliding window This type of window has the capability to slide in order to open.

slope Slope refers to how fast the roof rises from the horizontal.

soffit A covering for the underside of the overhang of a roof.

soil stack A soil stack is the ventilation pipe that comes out of a roof to allow the plumbing to operate properly inside the house. It is usually made of a soil pipe (cast iron). In most modern housing, the soil stack is made of plastic.

soleplate A soleplate is a 2 × 4 or 2 × 6 used to support studs in a horizontal position. It is placed against the flooring and nailed into position onto the subflooring.

span The span of a roof is the distance over the wall plates.

spreader Special braces used across the top of concrete forms are called spreaders.

square This term refers to a shingle-covering area. A square consists of 100 square feet of area covered by shingles.

square butt strip This refers to shingles for roofing purposes that were made square in shape but produced in strips for ease in application.

staging This is the planking for ladder jacks. It holds the roofer or shingles.

stain Stain is a paint-like material that imparts a color to wood. It is usually finished by a clear coating of shellac, varnish or satinlac, or brush lacquer.

stapler This device is used to place wire staples into a roof's tar paper to hold it in place while the shingles are applied.

steel square The steel square consists of two parts—the blade and tongue or the body.

stepped footing This is footing that may be located on a number of levels.

stile A stile is an upright framing member in a panel door.

stool The flat shelf that rims the bottom of a window frame on the inside of a wall.

stop This applies to a door. It is the strip on the door frame that stops the door from swinging past the center of the frame.

storm door A storm door is designed to fit over the outside doors of a house. It may be made of wood, metal, or plastic, and it adds to the insulation qualities of a house. A storm door may be all glass, all screen, or a combination of both. It may be used in summer, winter, or both.

storm window Older windows have storms fitted on the outside. The storms consist of another window that fits over the existing window. The purpose is to trap air that will become an insulating layer to prevent heat transfer during the winter. Newer windows have thermopanes, or two panes mounted in the same frame.

stress skin panels These are large prebuilt panels used as walls, floors, and roof decks. They are built in a factory and hauled to the building site.

strike-off After tamping, concrete is leveled with a long board called a strike-off.

strike plate This is mounted on the door frame. The lock plunger goes into the hole in the strike plate and rests against the metal part of the plate to hold the door secure against the door stop.

striker This refers to the strike plate. The striker is the movable part of the lock that retracts into the door once it hits the striker plate.

stringer A carriage is also called a stringer.

strip flooring Wooden strip flooring is nothing more than the wooden strips that are applied perpendicular to the joists.

strongbacks Strongbacks are braces used across ceiling joints. They help align, space, and strengthen joists for drywall installation.

stucco Stucco is a type of finish used on the outside of a building. It is a masonry finish that can be put on over any type of wall. It is applied over a wire mesh nailed to the wall.

stud This refers to the vertical boards (usually 2×4 or 2×6) that make up the walls of a building.

stump A stump is that part of a tree which is left after the top has been cut and removed. The stump remains in the ground.

subfloor The subfloor is a platform that supports the rest of the structure. It is also referred to as the underlayment.

sump pump This refers to a pump mounted in a sump or well created to catch water from around the foundation of a house. The pump takes water from the well and lifts it to the grade level or to a storm sewer nearby.

surveying Surveying means taking in the total scene. In this case, it refers to checking out the plot plan and the relationship of the proposed building with others located within eyesight.

suspended beams False beams may be used to lower a ceiling like a grid system; these beams are suspended. They use screw eyes attached to the existing ceiling joists.

sway brace A sway brace is a piece of 2×4 or similar material used to temporarily brace a wall from the wind until it is secured.

swinging door A swinging door is mounted so that it will swing into or out of either of two rooms.

table saw A table saw is electrically powered, with a motor-mounted saw blade supported by a table that allows the wood to be pushed over the table into the cutting blade.

tail The tail is the portion of a rafter that extends beyond the outside edge of the plate.

tail joist This is a short beam or joist supported in a wall on one end and by a header on the other.

tamp To tamp means to pack tightly. The term usually refers to making sand tightly packed or making concrete mixed properly in a form to get rid of air pockets that may form with a quick pouring.

taping and bedding This refers to drywall finishing. Taping is the application of a strip of specially prepared tape to drywall joints; bedding means embedding the tape in the joint to increase its structural strength.

team A team is a group of people working together.

terrazzo This refers to two layers of flooring made from concrete and marble chips. The surface is ground to a very smooth finish.

texture paint This is a very thick paint that will leave a texture or pattern. It can be shaped to cover cracked ceilings or walls or beautify an otherwise dull room.

thermal ceilings These are ceilings that are insulated with batts of insulation to prevent loss of heat or cooling. They are usually drop ceilings.

tie A tie is a soft metal wire that is twisted around a rebar or reinforcement rod and chair to hold the rod in place till concrete is poured.

tin snips This refers to a pair of scissors-type cutters used to cut flashing and some types of shingles.

tongue-and-groove Roof decking may have a groove cut in one side and tongue led in the other edge of the piece of wood so that the two adjacent pieces will fit together tightly.

track This refers to the metal support system that allows the bifold and other hung doors to move from closed to open.

transit-mix truck In some parts of the country, this is called a Redi-Mix truck. It mixes the concrete on its way from the source of materials to the building site where it is needed.

traverse rod This is another name for a curtain rod.

tread The part of a stair on which people step is the tread.

trestle jack Trestle jacks are used for low platforms both inside and outside. A ledger, made of 2×4 lumber, is used to connect two trestle jacks. Platform boards are then placed across the two ledgers.

trimmer A trimmer is a piece of lumber, usually a 2×4, that is shorter than the stud or rafter but is used to fill in where the longer piece would have been normally spaced except for the window or door opening or some other opening in the roof or floor or wall.

trowel A trowel is a tool used to work with concrete or mortar.

truss This is a type of support for a building roof that is prefabricated and delivered to the site. The W and King trusses are the most popular.

try square A try square can be used to mark small pieces for cutting. If one edge is straight and the handle part of the square is placed against this straightedge, the blade can be used to mark the wood perpendicular to the edge.

underlayment This is also referred to as the subfloor. It is used to support the rest of the building. The term may also refer to the sheathing used to cover rafters and serve as a base for roofing.

unfaced insulation This type of insulation does not have a facing or plastic membrane over one side of it. It has to be placed on top of existing insulation. If used in a wall, it has to be covered by a plastic film to ensure a vapor barrier.

union A union is a group of people with the same interests and with proper representation for achieving their objectives.

utilities Utilities are the things needed to make a house a home. They include electricity, water, gas, and phone service. Sewage is a utility that is usually determined to be part of the water installation.

utility knife This type of knife is used to cut the underlayment or the shingles to make sure they fit the area assigned to them. It is also used to cut the saturated felt paper over a deck.

valley This refers to the area of a roof where two sections come together and form a depression.

valley rafters A valley rafter is a rafter which extends diagonally from the plate to the ridge at the line of intersection of two roof surfaces.

vapor barrier This is the same as a moisture barrier.

veneer A veneer is a thin layer or sheet of wood.

vent A vent is usually a hole in the eaves or soffit to allow the circulation of air over an insulated ceiling. It is usually covered with a piece of metal or screen.

ventilation Ventilation refers to the exchange of air, or the movement of air through a building. This may be done naturally through doors and windows or mechanically by motor-driven fans.

vernier This is a fine adjustment on a transit that allows for greater accuracy in the device when it is used for layout or leveling jobs at a construction site.

vinyl Vinyl is a plastic material. The term usually refers to polyvinyl chloride. It is used in weatherstripping and in making floor tile.

vinyl-asbestos tile This is a floor covering made from vinyl with an asbestos filling.

water hammer The pounding sound produced when the water is turned off quickly. It can be reduced by placing short pieces of pipe, capped off at one end, above the most likely causes of quick turnoffs, usually the dishwasher and clothes washing machine.

water tables This refers to the amount of water that is present in any area. The moisture may be from rain or snow.

weatherstripping This refers to adding insulating material around windows and doors to prevent the heat loss associated with cracks.

winder This refers to the fan-shaped steps that allow the stairway to change direction without a landing.

window apron The window apron is the flat part of the interior trim of a window. It is located next to the wall and directly beneath the window stool.

window stool A window stool is the flat narrow shelf which forms the top member of the interior trim at the bottom of a window.

wrecking bar This tool has a number of names. It is used to pry boards loose or to extract nails. It is a specially treated steel bar that provides leverage.

woven This refers to a type of roofing. Woven valley-type shingling allows the two intersecting pieces of shingle to be woven into a pattern as they progress up the roof. The valley is not exposed to the weather but is covered by shingles.

zoning laws Zoning laws determine what type of structure can be placed in a given area. Most communities now have a master plan which recognizes residential, commercial, and industrial zones for building.

Index

ABOUT THE AUTHORS

Rex Miller is Professor Emeritus of Industrial Technology at State University College at Buffalo and has taught technical curriculum at the college level for more than 40 years. He is the co-author of the best-selling *Carpentry & Construction,* now in its fourth edition, and the author of more than 75 texts for vocational and industrial arts programs. He lives in Round Rock, Texas.

Mark R. Miller is Professor of Industrial Technology at The University of Texas in Tyler, Texas. He teaches construction courses for future middle managers in the trade. He is coauthor of several technical books, including the best-selling *Carpentry & Construction,* now in its fourth edition. He lives in Tyler.